D0145093

THE CRITICAL VILLAGER

Development aid is often ineffective and unsustainable. The scale of problems being faced by the Third World demands large-scale, replicable solutions but the high rate of failure in aid projects is often ascribed to inadequate consideration of local culture and conditions. Can demands for generalizable actions be reconciled with location-specific solutions?

The Critical Villager argues that community-based participatory research and 'transfer of technology' are not rival models of development but complementary components of effective aid. The eight practical principles for evaluation and action described call for students, development workers, policy makers and researchers to put themselves in the shoes of the intended beneficiaries of aid. *The Critical Villager* suggests that despite the wide range of cultures and circumstances there are certain constant principles underlying how people select new technologies and practices which can guide how aid interventions are designed.

Eric Dudley is a freelance rural development consultant and Director of Cambridge Architectural Research.

THE CRITICAL VILLAGER

VILLAGER

Beyond community participation

Eric Dudley

London and New York

First published 1993
by Routledge
11 New Fetter Lane, London EC4P 4EE

Simultaneously published in the USA and Canada
by Routledge
29 West 35th Street, New York, NY 10001

©1993 Eric Dudley
Typeset by J&L Composition Ltd, Filey, North Yorkshire
Printed and bound in Great Britain by
Biddles Ltd, Guildford and King's Lynn

British Library Cataloguing in Publication Data
A catalogue reference for this book is available from the British Library.

ISBN 0–415–07343–X Hbk
ISBN 0–415–07344–8 Pbk

Library of Congress Cataloguing in Publication Data
has been applied for.

ISBN 0–415–07343–X Hbk
ISBN 0–415–07344–8 Pbk

*To the countless under-paid,
under-resourced and under-respected
development field workers who,
against all the odds,
achieve the occasional success.*

CONTENTS

FIGURES AND TABLES

FIGURES

TABLES

PREFACE

Development aid is difficult. In this book I do not dwell on the statistics, international finance and grand policy of development. Rather, I try and deal with the concrete problems of the individual aid worker faced with the needs and desires of the individual villager or shanty town dweller; the men and the women who across the world are the true engines and instruments of social and technological change. My intention is to present some concrete principles for action which can be of use as much to the solitary field worker as to the policy makers of development institutions and the academic researcher.

In the evolution of the ideas which this book reflects I am indebted to many people. The first debt is a collective one to the many field workers in various countries who have given their time, patience and experience. In particular, my thanks go to my former colleagues in Ecuador, notably Diego Jordan and Bolivar Romero, while in Pakistan special thanks are also due to numerous individuals from the institutions of the Aga Khan Development Network. Thanks go to CAFOD for financing my work in Ecuador and to the Science and Engineering Research Council for supporting subsequent research. Particular thanks go to Jolyon Leslie, Stephen Rasmussen and Ane Haaland for the inspiration of their work. Also Robin Spence, Babar Mumtaz, Michael Parkes and Robert Chambers for their comments on earlier drafts. The greatest debt is to Catharine Forbes with whom many of the ideas and practices were developed over several years.

Eric Dudley
Cambridge, 1993

INTRODUCTION

The accumulated experience of aid and development appears to pull in two contradictory directions. On one side, the high rate of failure in aid projects is frequently ascribed to an inadequate consideration of local culture and conditions. The panaceas of some earlier aid programmes have been discredited as simplistic and culturally insensitive. In their place are calls for aid to be executed by multi-disciplinary teams who are expected to tailor their interventions to the unique physical and social circumstances of each and every location.

Set against this tendency is the inescapable fact that the problems which gave rise to the earlier broad-brush programmes have not gone away; indeed, many of them have got worse. The scale of the problems being faced by the Third World compared to the limited resources available demands widely replicable solutions. We appear to be faced with a choice between quality and quantity; a handful of culturally sensitive but irreproducable projects or actions which ignore and damage local cultures and which are not sustainable since they have been imposed from outside. Neither choice is satisfactory. The objective of this book is to contribute to the reconciliation of the simultaneous demands for generalizable actions and location-specific solutions.

AID

The objective of all aid is to help bring about change, whether in terms of how a seed is sown or to whose benefit political power is used. Unlike academic anthropological studies, aid demands an understanding of the contradictions involved in the benevolent exercise of power.

1

There are two types of aid: material and technical. Because material aid involves large amounts of money it tends to attract more attention both from donors and potential beneficiaries. In development manifestos there is generally only a token sentence or two, tagged on near the end, advocating technical aid. The actual nature of technical aid remains largely undefined. It seems to be assumed that either technical aid is not important or else that we know how to do it. My contention is that neither assumption is true.

Since nobody wants to make open-ended commitments to aid, donors of material aid tend to describe their work in terms of transitional 'pump-priming' assistance before 'take-off' into self-sufficiency. In this light, the long-term goal of all aid, whether material or technical, is to bring about permanent changes in the way in which people do things with their own resources and from their own initiative. Certain forms of material aid have become well established, such as the provision of credit and infrastructure. However, widespread failures of credit programmes are reflected by a growing preoccupation in project proposals with mechanisms of debt recovery. Similarly, more and more infrastructure projects seem to be concerned not with the installation of new infrastructure so much as the renovation of earlier irrigation channels, health centres, sewers and schools which were themselves the products of aid only a few years before.

Achieving superficial and short-lived success in a material aid project tells us nothing about how to do technical aid. In contrast, if we could learn how to bring about advances solely through technical aid it is likely that we could also understand how to make material aid more lasting in its effects. One villager who adopts a technology at his own initiative and with his own resources is more significant in the process of change than the many who have been paid by outsiders to use the technology. An understanding of technical aid is important not as an appendage to material aid but as its foundation.

For some people the neglect of technical aid has not been an oversight but a deliberate decision; they question whether technical aid is a valid aid strategy. These critics, from both left and right on the political spectrum, believe that development is about access to and control over material goods. They further believe that people generally know what they want and how, if only they have sufficient resources, to get it. In this view, technical aid is seen, from within the Third World and without, as a product of a patronizing attitude on the part of an outsider who believes that he is better able to say what is good for someone than that person can him- or herself. The critics consider the attitude of technical aid workers to be both mistaken and racist.

In my view these criticisms have a certain validity, but there is still a place for technical aid. Quite apart from the need to understand the processes involved as a basis for material aid there appear to be three reasons for valuing technical aid in its own right:

- The nature of need is changing. If needs and aspirations were static it would be arrogant to suppose that identification of the most appropriate techniques could be achieved through the intervention of outsiders. However, the nature of the problems being faced in Third World villages is changing at an unprecedented rate. Indigenous technical knowledge is often inadequate for the novel problems presented by ecological degradation, changing aspirations and new social institutions. In such circumstances, there is a case for external expertise in the new technologies and in the processes of technical and institutional change commonly found in societies in transition.
- Some useful technical knowledge cannot be readily deduced from the kind of knowledge current in many Third World villages. This is particularly evident with regard to health care where such things as the relationship between hygiene and health is not apparent without the benefit of modern science. Recognizing the value of indigenous knowledge does not automatically negate the value of information from outside.
- Material aid is inadequate and, for the foreseeable future, is likely to remain so. The overwhelming majority of rural development actions are initiated, implemented, and paid for by the rural poor. The disparity between the material resources required and those which are available suggests that the standard of living of the majority will not be improved through material aid alone. If there are to be significant improvements in living conditions they will need to come about largely through better use of existing resources.

Material aid has the great advantage of intelligibility; the donors understand what they are giving and, notwithstanding problems of tied aid and re-payment conditions, the beneficiaries understand what they are receiving. Despite all the evidence of long-term failures, money can always buy gratifying and politically expedient short-term success. A key measure of a donor agency's success is whether it has succeeded in spending all of its allotted money within the financial year. Other measures of an investment's long-term efficacy are neither perceived as pressing nor easily determined. In recipient countries, the principal beneficiaries of material aid frequently appear to be officials and consultants, the benefits taking the form of imported vehicles, salaries and fees.

This cynical but familiar picture is simplistic and uncharitable. There are plenty of honest and intelligent people trying to apply material aid in effective and innovative ways, but this is not the point. The increasing sophistication and complexity of material aid, whether in the form of ingenious mechanisms for credit, transference of debt, or food for work, simply reflects the comprehensible nature of material resources. They are amenable to refinement, discussion, and policy because they are intelligible. People understand money; they do not understand technical aid.

The aid community has tried to identify indicators of technical aid activity. The number of booklets distributed and the number of people 'trained' on courses are taken as tangible indices of technical aid. Such quantitative measures give no indication as to the efficacy of the aid projects, they just indicate that the intervening institution is doing something, for good or for bad. Technical aid continues to be largely undiscussed in policy statements because it is not understood.

In the 1970s the 'Appropriate Technology movement' appeared to offer a simple, intelligible and attractive model for technical aid: 'The best aid to give is intellectual aid, a gift of useful knowledge. A gift of knowledge is infinitely preferable to a gift of material things' (Schumacher 1973: 192). The name of Schumacher, through his influential book *Small is Beautiful* (1973) and the creation of the Intermediate Technology Development Group (ITDG) in 1965, is inextricably linked with the terms Appropriate Technology and Intermediate Technology. Schumacher believed that 'the necessary knowledge, by and large, exists', and that 'the task of aid [is] primarily one of supplying relevant knowledge, experience, know-how etc. – that is to say, intellectual rather than material goods' (Schumacher 1973: 191,193). Schumacher's initial concern was not the development of new technologies but the transfer or dissemination of existing tried and tested knowledge. It is significant that the first act of ITDG was the publication of a catalogue of commercially available agricultural tools (ITDG 1967).

4

The notion of Appropriate Technology came under attack from a number of directions. From a Marxist standpoint, the Appropriate Technology movement was stigmatized as a means of maintaining the status quo:

> In advanced capitalist and Third World countries alike inter-mediate technology and self-help philosophies are put forward as a solution: build your own house, grow your own food, bicycle to work, become an artisan and so on. To those in the Third World who have done all these things and who are still rarely far from starvation, such appeals to be more self-reliant must strike them as being a rather curious form of radicalism.
>
> (Burgess 1982: 92)

In an allied view, Appropriate Technologies are seen, particularly by people themselves in the Third World, as 'second-rate technologies'. This description is rejected by many exponents of Appropriate Technology (eg Schumacher 1973: 150–1; Turner 1976: 108). Others accept that such technologies are indeed second rate: 'Appropriate Technology is proposed for the urban ghettos, for the rural poor, for Navajo Indians, or for Newfoundland fisherman. It is precisely as a "poor people's technology" that it is viewed' (Rybczynski 1980: 86). In this view, the people of the Third World usually see Appropriate Technologies as, at best, stepping stones in the process to modernization (Rybczynski 1980: 67–82).

Perhaps the most attractive aspect of the Appropriate Technology theory has also been its greatest pitfall: the lure of the 'technical fix'. The idea that a clever, cheap, device will solve a problem of development can be very attractive to all sorts of people involved with development. The big financial institutions are attracted by a tangible solution with a price tag which allows people to 'stand on their own two feet'; the aid agencies are presented with the prospect of by-passing state bureaucracies and corruption through the use of village-level projects; the engineers and technologists see an opportunity to contribute to the greater good by applying the skills for which they have been trained and which they enjoy; the environmentalists are attracted by the use of local materials and energy resources; community workers have seen opportunities for local initiatives which hold out the hope for liberating people from dependencies which otherwise seem beyond their control; and the charity-giving public, frustrated by the evident failures of aid, is attracted by photographs of people benefiting from an intelligible and, apparently, demonstrably good idea. In practice, enthusiasm for ingenious human-scale technologies may result in technically innovative but, nevertheless, inappropriate solutions as witnessed by the plethora of solar panels and windmills in the 1970s and the theoretically fuel-efficient but often impractical cook-stoves of the 1980s.

When the idea of Appropriate Technology was new, the debate necessarily remained largely on a theoretical level. Now, there is a body of experience of Appropriate Technology projects much of which has proven to be negative. Among many aid agencies Appropriate Technology appears to be regarded as a blown myth. The great attraction of Schumacher's concept was the belief that a good idea will spread of its own accord. Notwithstanding the argument for 'trickle-down' benefits, material aid generally only reaches the immediate beneficiaries of the project whereas 'intellectual aid' does have the potential to self-propagate. In practice, with few exceptions, the benefits of Appropriate Technology projects have proven to be as fenced in by a project's geographical boundaries as those of any other type of aid project. Similarly, the apparent success is often limited to a project's temporal boundaries. In the same way that material aid can often only sustain success for the duration of the subsidy, so too are Appropriate Technologies frequently dependent on the presence of an enthusiastic champion; the field worker who ensures technical back-up and cajoles the 'beneficiaries' into using the product. Marilyn Carr, while Head of Policy Planning at the ITDG, described this inability to get beyond the confines of the project as 'a crisis for the Appropriate Technology movement' (Carr 1985: 369).

From an early stage, some critics held that it was fundamentally mistaken to think of Appropriate Technology in terms of the transfer of technological artifacts. Rather, the Appropriate Technology movement should concern itself with a consideration of the process by which Third World countries can develop their indigenous innovative capacity (Jéquier 1976: 25). Among theorists there seems to be widespread agreement on the correctness of this objective, but the means by which it can be reached remain elusive. Institutions and governments need demonstrable and quantifiable results to show the electorate, the bankers and the charity-supporting public. Faced with this pragmatic demand, along with the day-to-day realities of budgets and timetables, development institutions often continue to resort to finite projects and tangible technologies.

COMMUNITY PARTICIPATION: US AND THEM

Shifting focus onto the indigenous innovation process inevitably leads to a preoccupation with community participation. Participation used to be the rallying cry of the radicals; its presence is now effectively obligatory in all policy documents and project proposals from the international donors and implementing agencies. Community participation may have won the war of words but, beyond the rhetoric, its success is less evident. Part of the problem is clearly political. True participation is a threat to powerful vested interests: 'participation is applauded; encounter is not' (Chaufan 1983: 88). Yet, the difficulty can not all be ascribed to conspiracy theory; even where the commitment to participation is genuine, there are many different ideas about what that really means.

The most fundamental split is between those who see participation as a means to an end and those who advocate it as an end in itself. As a goal in itself, community participation appears necessary to 'stimulate individual and social well-being' (Turner and Fichter 1972). This view has been related to theories of neighbourhood or workplace democracy which have emerged from dissatisfaction with the democratic nation-state (Midgley 1986: 15; Schumacher 1973: 235–85). Those who consider participation as a goal in itself are characterized by one author as romantics in search of 'community lost' (Rybczynski 1980: 26). Although discussion of the spiritual benefits of participation rarely finds its way into project documents it seems to be an unstated ethic underlying a lot of aid work. It is common to find middle-class urban aid workers, children of the mobile society, lecturing to close-knit communities of villagers about the need to work together.

It is more usual for community participation to be considered as a tool for carrying out a task, whether political or physical. Its most obvious use as a political tool is to bring people together to lobby the state to provide services. Or, expressed more radically, 'unity among the oppressed' is considered as a necessary prerequisite to liberation (Freire 1972: 172–3). Since it is hard to challenge the inherent goodness of community participation it has become a double-edged tool sometimes used to justify the state's evasion of its own responsibilities. The state may transfer responsibility to 'the community' or the 'voluntary sector' for services which previously have been considered the duty of the state (Midgley 1986: 41).

Government and international agencies generally describe community participation less controversially as a method to accomplish physical tasks both more cost-effectively and with a greater like-lihood of sustainability. With their realization that the scale of the problems is too great for governments to handle by conventional means, participation has become an economic necessity. Yet, despite a commitment to the idea of community participation and self-help, there is bewilderment as to how governments can support it on a sufficient scale.

The three functions of community participation as goal, political tool and physical tool are not incompatible but these three categories are frequently confused. A public commitment to participation often conceals doubts as to how and why it might apply. Although it has been top of the agenda for twenty years it is still far from clear what community participation is, how it comes about, and what it is actually for.

A fourth function of the term community participation is for development workers to persuade themselves that the activity in which they are involved is not paternalistic. All aid is paternalistic.

> However much the rhetoric changes to 'participation', 'participatory research', 'community involvement' and the like, at the end of the day there is still an outsider seeking to change things. Marxist, socialist, capitalist, Muslim, Christian, Hindu, Buddhist, humanist, male, female, young, old, national, foreigner, black, brown, white – *who* the outsider is may change but the relation is the same. A stronger person wants to change things for a person who is weaker. From this paternal trap there is no complete escape.
>
> (Chambers 1983: 141)

Aid is predicated on the existence of outsiders with resources – and resources mean power. The selfish intervener can use that power to manipulate events, such as trade deals, to his own advantage. The genuinely philanthropic intervener is faced with the problem of how to exercise that power in the most equitable, just, and effective manner.

The language of community participation has tended to obscure the fact that there is always an 'us' and a 'them' – there are those who are trying to address the needs of others and there are those who are the target of philanthropy. The borderline between the two groups is not fixed. In some circumstances 'us' means the governments of the handful of wealthy industrial nations and 'them' is everybody else. But, in other circumstances, 'us' may include local field workers who are drawn from, and live in, the villages which are the target of aid. Similarly 'them' sometimes refers solely to government ministries, at other times householders, and sometimes more specific groups such as homeless street children.

Talk about 'us' and 'them' is not about value judgements. It is not being suggested that 'we' are somehow better than 'they' are. It is no more a value judgement than when a shopkeeper refers to his customers as the general public. The public are the target for the service which he offers. In the context of aid 'we' refers to that group engaged in the business of offering development aid. Simply using the magic words 'community participation' will not result in us being accepted as one of them.

The primary concern of this book is with aid directed to the villagers and shanty town dwellers of the Third World. One could devote pages to trying to define and defend the term 'Third World'. Here it is used simply as a widely understood shorthand for those countries which, in material terms, are less well off. These targets of aid are here referred to as the 'intended beneficiaries'. This phrase is chosen in preference to something more neutral like the community, farmers, villagers, or the people in order to emphasize both that:

- The object is not to produce research papers but to actually benefit people; and
- Simply carrying out an aid programme does not guarantee that the intended targets will benefit. It only expresses an intention. Much aid does no good and some does harm.

The other partners in the aid equation – 'us' – are described as 'interveners'. Everyone whose business is development aid, from the field worker to the international banker, is intervening in someone else's process of development. By choosing to help one group rather than another or one activity in preference to another these actors in the development process are intervening to bring about change. The interveners may be foreign or national, commercial consultants or volunteers, religious or secular, self-appointed or elected. They are all intervening in other people's lives in the name of the aid process.

THE INTERVENERS

Academic studies, political rhetoric, and consultants' reports frequently call for aid interveners to collaborate and to have an integrated, holistic and multi-disciplinary approach. There are two obstacles to achieving this aim. First, this approach assumes that intervening institutions are all trying to achieve the same thing, and second it presupposes that we know how to work in a multi-disciplinary manner.

What makes Third World development both an interesting and an immensely difficult field is that all facets of life are involved. It deals with the growth and change of a civilization in all its aspects; economy, politics, technology and culture. If political ideology is conflictive in the developed world, with its long-established social structures, it is hardly surprising to find energetic differences of viewpoint when the whole future form of an emergent modern state is open to influence. Non-Governmental Organizations (NGOs) are often portrayed in the saintly light of latter-day secular mission-aries doing good among the poor. In practice, many local NGOs are established either to promote a particular political or religious view or out of anger at the perceived inability of government and other NGOs to meet the legitimate demands of marginalized and oppressed people. Lack of collaboration among these organizations may not reflect oversight or inefficiency so much as an active hostility.

It is only the large well financed projects which are likely to be able to support multi-disciplinary policy-making teams. But the creation of such a team is not, in itself, the answer to the problems of complexity. 'Multi-disciplinary team' can easily become a paper-term which screens the reality of a disparate set of individuals working as they always did, in isolated professional pigeon holes. The technologist may recognize that culture is important yet not know how to make practical use of the anthropologist's reports, however interesting they may seem. Equally, the anthropologist, though anxious to make a useful contribution, may not know how to ask the questions that will produce useful answers for the implementer. The result is the retreat to safe, discrete, professional patches and the production of a multi-disciplinary set of mono-disciplinary papers. The problem of synthesizing the results is tacitly passed on to some other undefined actor in the process. The impression formed by some, including myself, is that in the overwhelming majority of those cases where the synthesis does occur it happens at the level of the individual field worker or the exceptionally dynamic community leader:

The burden [is] on the field worker in a particular project, who is expected to have multiple skills to achieve an integration at the grass roots which does not exist anywhere else along the line. Remarkably, many field workers do have these multiple skills, but . . . particularly in the case of volunteers, there is a very real danger of overburdening one individual with too many different tasks resulting in nothing getting done properly.

(Chaufan 1983: 10)

The field worker has been largely ignored in the literature on development aid. While researchers have stressed that the community is not an homogeneous whole, the intervening agency has still been generally treated as a single entity. The term 'field worker' is an umbrella expression which covers a large variety of people. A few are overseas volunteers who themselves vary from school leavers to highly experienced professionals, many are college or high school graduates from the local towns, while others are recruited directly from the villages which they serve.

The successful field worker who is capable of stimulating and supporting well-rounded, community-based, integrated rural development has to be a kind of renaissance generalist. Overstretched and under-resourced, the field worker must juggle the issues and strike pragmatic compromises between policies which tend to come to the field in the form of contradictory messages. Policy may decree that community participation, self-determination, and village-level democracy are essential while at the same time holding that ecological considerations and the use of indigenous technologies are paramount. Policy may demand the emancipation of women while insisting on respect for traditional cultural mores and institutions. The field worker, faced with democratic and vociferous community demands for pesticides and tractors and the frosty rejection of birth control measures, is left with the task of resolving the unresolvable while keeping his or her employers happy.

In theoretical discussion, people will readily agree that failures are an important part of the learning process. When considering their own specific projects, interveners at all levels in the process have an interest in presenting a picture of success. Meaningful evaluation and institutional learning are obstructed by a conspiracy of success. Success is rewarded while failure, however potentially informative, is not. The field worker on a short-term renewable contract needs to meet targets of numbers of trees planted or latrine squat-slabs distributed. If the trees die or the squat-slabs lie unused the information is liable to be studiously ignored or optimistically massaged. The policy makers, watchful of future funding, succumb to the pressure, imagined or otherwise, to report to their donors that objectives were met. The donors, aware of the stiff competition in the compassion industry, need to report a cosy scene of gratitude and money well spent. The knowledge of the nature of failure, the very information which could allow intervention policy to be improved, is lost.

Criteria used by the various interveners may differ wildly. Members of the professional policy-making team have their own angles as do the field workers and the donors. A group of individuals can only work as a cohesive team when they share a single definition of their shared problems. Given the complex and all-embracing nature of development, it is not reasonable to expect everybody or indeed anybody in the aid process to be fully informed on all aspects of the problems. What seems to be required is a shared intellectual framework around which developmental problems may be analysed and aid interventions designed, implemented, and evaluated; a framework which is loose enough to be broadly acceptable and understood, yet specific enough to point to concrete actions rather than empty theoretical debate. The framework needs to enable the village-level field worker to engage in a mutually useful exchange of ideas with the high-flying, although possibly tunnel-visioned, university trained professional.

THE INTENDED BENEFICIARIES

The strategies of aid institutions and their criteria of appropriateness are inevitably dictated by the ideological perspective of the interveners. In constructing a common framework for action it would, I suggest, be useful to focus not on the ideology of change but on the mechanics and nature of change. In seeking such a framework I have started from two trivial observations:

- **The goal of aid is change** Technical aid is about introducing changes in both technologies and the processes of technological innovation.
- **Change is going on all the time** The overwhelming majority

of change is initiated and paid for by the users. Anybody wishing to introduce technology change should start by examining the indigenous processes through which the intended beneficiaries of aid make their own technology choices.

From these observations emerges the central proposition of this book: *Aid should be change-like.* Technical aid interventions are more likely to be successful if they share characteristics with the indigenous processes of technology change.

The traditional image of the target of charity has been that of the passive and grateful recipient of whatever crumbs the philanthropic benefactor has been inclined to bestow. Schumacher's model of technical aid appears to have been predicated on a rather similar image of people waiting to be told an answer which when it came along they would enthusiastically embrace. In this picture, all that seems to be necessary is for the technologist to identify solutions and to reveal the good news. In practice, this model has proven to be both inaccurate and incomplete. The villager and shanty-town dweller are neither passive nor willing to accept an outsider's proffered solution as unquestionably correct. They are not empty vessels waiting to be filled, but sceptical scientists and shrewd consumers energetically pursuing the best available solution to their problem of the moment. Unless the process is distorted by material incentives, the aid intervener is generally just one source among many which the intended beneficiary calls upon for information and ideas.

The widespread failure of projects has helped to stimulate awareness of the importance of the cultural aspects of technology and recognition of the potential value of indigenous technical knowledge (see for example Brokensha *et al.* 1980). During the 1980s the word 'anthropologist' became one of the key signs which marked a project proposal as 'progressive'. The studies of specific cultures were primarily intended to identify 'felt needs' as well as the reactions and possible obstacles to proposed new technologies; in other words, they were to inform what was essentially still an externally driven process. The villager was a patient with needs to be attended to.

More recently, the emphasis switched to focus on the activities of the villager as innovator and prime instigator of change (for instance, Chambers *et al.* 1989). The outside researchers and aid workers are coming to see themselves as 'enablers' and 'resources' to an indigenous process. Even the use of these titles is, perhaps, encouraging an inflated self-image for the intervener. When the distortions of material aid are stripped away, the best an intervener can hope for is, in some small way, to assist and perhaps broaden the choice-process of the intended beneficiary. In order to do this, the outsider needs to consider the motivation of the villager or shanty town dweller and anticipate the kinds of question which he or she will need to answer before undertaking change.

Some would suggest that the intervener has nothing to contribute regarding informing people of their own motivations. Conversely, the proponents of 'consciousness raising' techniques as developed by Paulo Freire argue that the intervener has a role in helping the beneficiary to recognize the true nature of his or her circumstances. Whichever view is taken, it is clear that an understanding of both motivations and circumstances is vital for the intervener before presenting his ideas to the scrutiny of the critical villager.

PRINCIPLES

The answers people find to their questions clearly vary enormously with place and time, but I suggest there are common elements in the behaviour of the critical villager which may be reflected in a set of replicable principles. These general principles may be used as the shared framework for discussing and generating location-specific solutions both for technologies and the dissemination process. I suggest that across a broad range of circumstances people of all types are asking three basic questions of new ideas:

- **Does it make sense?** Is the idea *reasonable* in terms of the intended beneficiary's own rationale?
- **What is it?** Can the idea even be *recognized* – does it have a name and are its limits clearly defined?
- **Is it worthy of me?** Is the idea *respectable* – is it something which 'people like us' do?

These three preoccupations can be broken down into eight more specific principles. The following eight chapters are each devoted to one such principle. Unlike the kinds of list which attempt to define what is 'appropriate', the eight principles do not describe ideological 'oughts'; rather, they point to characteristics which are liable to make a technology more appropriate-able; for good or ill.

The proposed principles do not describe a comprehensive recipe for technical aid. They are little more than an attempt to record the common-sense knowledge in the mind of the competent aid worker. As such, the principles may be criticized as rather empty, yet the evidence seems to suggest that many people who design and implement development projects lose sight of common sense. By using this book to give the obvious a name, my intention is to identify reference points which, across a range of cultural contexts, anchor a project within a common frame of discussion and evaluation. The most important purpose of the principles is to generate questions which can produce useful action-orientated answers for all the actors in the intervention process.

In this book, with a few notable exceptions, rather than relying on secondary sources I have chosen to use examples either from my own work or from projects which I have visited. Since the topic of development is so vast, the experience of any author is going to be limited to certain sub-themes. I am male, white, and British; my training is as an architect and much of my field experience has been with rural housing and sanitation working from a small rural development centre called Centro Sinchaguasin in the Ecuadorian Andes. Inevitably, many of the examples and the arguments which I use in this book reflect this particular experience. However, the reader is invited to look beyond the bricks and mortar of the specific examples to see the principles which are being illustrated. Subsequent exposure to health and agriculture projects in Asia, Africa and other parts of Latin America convinces me that the principles introduced here may be usefully applied in discussing, designing and evaluating the process of aid whatever its content and wherever its location.

Part I

REASONABLE

For a new idea to be adopted it must make sense in terms of the intended user's own rationale. It must be clear what aspect of the idea is new within the context of existing knowledge and it must fit into the understood social fabric of responsibilities and skills. In order to understand what other people will consider reasonable it is necessary to find ways of learning the criteria, knowledge and priorities of others.

Figure 1 Protecting an earth wall
The effective life of a traditional earth wall can be significantly extended by applying a narrow strip of fibre-reinforced cement mortar to its top to protect it from the direct impact of rain. The technique is readily understood and can be easily added to the existing repertoire of skills.

1

THE BIG IDEA

> This century has been by far the most remarkable, in the
> intellectual history of the world, for its great progress in
> scientific discovery and invention. But in the midst of all the
> beneficial inventions made during the period there is one
> which is wholly *evil* – I mean the water-closet.
>
> (Charles Richardson 1893)

Designers are trained to find physical solutions to problems. A
designer suggests that if a series of steps are taken and a set of
objects put together in a certain way then the given problem will
be solved. The inexperienced designer tends to assume that the
procedures which he has described will be executed precisely as
intended. In practice, all of the operations will be done more or less
inaccurately and some may not be done at all. The designer, carried
away by the excitement of what is possible, can lose sight of what
can possibly or even probably go wrong. Maximum potential
performance can come to take priority over maximum tolerance to
error.

In the 1960s, some of the designers of pre-fabricated building
systems assumed that each piece would be made to the perfectly
correct size and placed in precisely the right place. In practice, many
of the buildings leaked or fell apart because the execution on
the building site did not match the geometric perfection of the
drawings. The scientists of the 'green revolution' designed hybrid
crops which could produce miraculous yields. When used on the
tiny farms of Third World smallholders without the intended
fertilizers, pesticides, irrigation and storage the results were often
disappointing and sometimes disastrous. More recently, much
effort and ingenuity has been invested in the design of improved

cook-stoves. The fuel savings achieved from mud stoves in which well dried firewood of a given diameter was burnt were remarkable, in some cases. When the stoves were reproduced in the villages without sufficient attention to the precise form of the combustion chambers and using the damp, irregular, firewood actually available, the savings achieved were commonly reduced to nil.

Much of the work on Appropriate Technologies seems to have suffered from a form of evangelical barrel vision. The results which could be achieved were assumed to be those that would be achieved. Since, under controlled conditions, the desired results could be produced, designers have been tempted to consider their designs to be demonstrably sound. Failures in the practice are attributed to conservatism, ignorance, carelessness or the mysterious vagaries of 'cultural factors'. For the professional designers – the university-trained engineers, architects and scientists, whether expatriate or national – such elements of failure are often considered to be outside their sphere of expertise. In their view, the technology works and resistance to the idea and shoddy workmanship must be dealt with by others through education. But such a view is frequently used as an excuse to shrug off responsibility for failures which could have been anticipated. The competent designer needs to ask two questions:

- How may the problem be solved?
- How may the proposed solution fail?

The second question is often neglected. All technologies can fail but some have a greater tolerance to less than perfect implementation than others.

HOW TECHNOLOGIES FAIL

The conditions experienced in Third World villages are generally different to those found in laboratories and demonstration projects. The facilities and the assumptions common in the laboratory may be absent in the village. University researchers in Peru experimented with various means of reinforcing rammed earth walls against earthquakes. Rammed earth is a technique in which, rather than building a wall with pre-formed bricks, moist earth is pounded with a wooden pestle into a large mould which is placed in the final position of the wall. The researchers proposed the use of vertical wooden poles cast into the foundations which pass up through the wall, the earth being rammed in around the poles. In laboratory tests, these poles were shown to strengthen the walls significantly.

The ideal earth for construction has a broad range of particle sizes. If the particles are predominantly large the finished wall will be weak and crumbly whereas if there are too many fine clay particles the wall will tend to shrink and crack as it dries out. After an earthquake in Ecuador, the technique of reinforcing rammed earth with vertical poles was adopted by a reconstruction project. Whereas in the laboratory experiments the earth had been of an ideal consistency, in the areas affected by the Ecuador earthquake the soil was rich in clay. When the pole reinforcers were employed in the reconstruction projects the poles acted as foci, concentrating and amplifying the drying cracks. The poles, rather than strengthening the walls, were fragmenting them and making them weaker than a simple un-reinforced wall. The technology, considered in isolation was not in itself bad but, like the 'green revolution' crops and the 'improved' cook-stoves, it was only good in a certain context. In some other contexts it was positively harmful.

When a technology is adopted a number of messages need to be understood and accepted. The number and content of the messages will vary according to the technology and the context. If all the messages are accepted and acted upon then the technology has been appropriated; if none are accepted then the technology is rejected and the intervention is a failure. This much is clear. Less clear are the many cases in which some of the messages associated with a technology are adopted while others are rejected. The intervener is left wondering whether the intervention was a success or a failure.

There is a need for an approach which allows the intervener to distinguish between messages which are vital to the successful functioning of the technology and others which are merely desirable refinements. In any situation, we can describe a new technology as having, potentially, up to four types of message associated with it:

- **The Big Idea** The single indivisible message which if not acted on would render the technology unrecognizable to all concerned.
- **Background deficit** Essential background knowledge which the intended user currently lacks but needs in order to be able to test and implement the Big Idea. The content of the background deficit is specific to a particular context since it is dependent on the initial state of the villager's background knowledge.
- **Essential practice** Messages which are not necessary to implement the Big Idea but which nevertheless are considered by the

outside intervener to be vital if the technology is to result in a significant net improvement in the intended beneficiary's situation. The essential practice is defined by the perspective of the intervener as much as by the nature of the technology itself.

● **Good practice** The 'getting the best out of your new . . .' type of information.

It is hard to know what we know. It is also difficult to realize what other people do not know. Sociologists refer to 'the world taken for granted' and the 'mind-set' against which an individual will judge new situations (Schutz 1964: 229). In Ecuador, a colleague was demonstrating an improved way of making an opening window during a housing improvement course, and one of the participants was overheard to say, 'I like a window in a room, it lets in light'. Traditional housing in the mountains has generally been dark and windowless due to the cold nights and the expense of glass. With changing patterns of living and rising aspirations windows were becoming a relevant technology. To this participant on the course, the concept of 'window' was still being absorbed and its costs and benefits established. It may well have been that an improved construction for a window was not an idea he was going to be ready to engage with until the deficiencies of conventional windows had been absorbed.

With some Appropriate Technologies the background knowledge which underlies large-scale factory production is being projected onto the village context. Village-made fibre-cement roofing tiles operate on the same technological principle as factory-made asbestos-cement roofing sheets. Natural fibres such as sisal or coconut fibre may be used to provide reinforcement to thin sheets of concrete in the same way that asbestos fibres can. Asbestos-cement roofing sheets are made in a process in which the mix of sand, cement, asbestos-fibre and water are mechanically controlled. The mix is pressed and vibrated into moulds to produce a uniform reliable product. Fibre-cement roofing tiles are intended to be manufactured in small-scale, village-level workshops which are expected to adopt a complex package of interrelated technological principles:

● Concrete is made stronger by vibrating it before it sets.
● If the sand used to make the concrete is clean and has a good range of particle sizes the concrete will be stronger than otherwise.
● The strength of concrete is influenced by the water content of the mix. Too much water will make the concrete easy to pour but will weaken the final product.

- The water should be clean.
- Concrete will achieve maximum strength if it is well cured (kept moist for a period of time).
- Concrete may be strengthened against tension and bending stresses by mixing in fibres.
- If fibres are used they must be well separated.

If fibre-cement tiles are made in large workshops under skilled supervision the consumer is presented with a product which may be tested. The purchaser only has to understand the Big Idea that the tiles keep the rain out, he does not need to understand the manufacturing process. The Big Idea that 'you can make a thin concrete tile reinforced with fibres which keeps the rain out' is not a testable proposition without a prior acceptance of all of the above principles; they are vital elements of background knowledge which must be understood and acted upon for the technology to work. A demonstration roof may be used to prove the efficacy of the product but not the validity of the above statements. The list is not definitive; there may be other relevant principles missing from somebody else's current knowledge. For some hypothetical isolated community, the fact that the contents of a cement bag mixed with sand and water will set to form a rock-like substance called concrete may be unknown. For other people, used to thatch roofs, concrete may be familiar while the idea of roof tiles is novel.

All of the statements form part of the existing background knowledge for trained engineers or architects. For them, the Big Idea is merely that 'fibre-cement roofing tiles may be made using natural fibres rather than asbestos fibres'. This is a statement which can be tested without any addition to their prior background knowledge. Once this new idea has been tested and found to work it forms part of the amended background knowledge against which can be tested the next Big Idea; such as, 'these tiles may be made using simple low-cost machinery'.

If a piece of essential background knowledge is not acted upon the technology fails in a way apparent to all. Any one of the background deficit messages for fibre-cement, if ignored, can result in tiles which break or let the rain through. Similar restrictions apply to more mainstream products. To someone used to living in circular thatched huts an asbestos-cement roofing sheet will be liable to fail as a roofing material. Large, stiff, rectangular roofing sheets only make sense in a context in which rectangular buildings are the norm and roof rafters are expected to be straight and line up with each other. Unless there is no background deficit the likelihood of successfully introducing a technology must be low. Identifying the background deficit could, however, suggest the nature of another intervention which will help to make up the deficit in preparing for a future programme. Today's Big Idea is tomorrow's background knowledge.

There are other messages regarded by development interveners as essential but which, when ignored, do not result in a technology which is perceived by the user to have failed. For instance, the use of chemical pesticides might be seen by an outsider to require the adoption of two messages:

- Spraying pesticides on your plants protects them from insects.
- It is necessary to wear a protective face mask to avoid serious and possibly fatal injury.

Similarly, the use of asbestos-cement roofing sheets can also be seen as requiring two key messages:

- Asbestos-cement roof sheets keep the rain out.
- In order to reduce the serious risk of cancer, it is necessary to take precautions such as wearing a face mask and dampening the sheets before cutting and drilling.

In practice, pesticides and asbestos-cement roofing sheets have both been widely adopted across the Third World on the basis of their respective first messages alone – their Big Ideas. The second message is not necessary in order to make the technology 'work' and so it is not part of the background deficit yet, to an outsider, the second message might reasonably be considered as vital.

The essential practice category of information only has meaning in the context of an outsider who is judging success against criteria which differs from that of the user. When a technology has an essential practice attached to it there exists the potential for partial appropriation. That is to say, the Big Idea may be adopted but practices which are essential for its safe implementation are not.

In considering pesticides and asbestos-cement, as well as products such as powdered milk and pharmaceuticals, the multinational corporations are accused of deliberately suppressing the dangers associated with their product. But there are other technologies, including some labelled Appropriate, which have been partially appropriated in ways which can cause actual harm to the users. In the case given above of the reinforcement to rammed earth there were two messages which the villagers in the project area needed to adopt:

- Use vertical poles to reinforce your rammed earth walls.
- Use soil with a low clay content.

The Big Idea was the first message. In some cases the second message constituted part of the background deficit. If the soil is very rich in clay the cracking caused by the poles can be so severe as to render the technology an obvious failure in the eyes of the users. In practice, the failure was often not so evident since the cracking was slight. An engineer might realize that the building has been weakened by the poles while the villager still believes he has done the sensible thing in using them.

A way of building a brick wall to be more resistant to earthquakes is to build the wall with buttresses. An earthquake reconstruction project in Ecuador promoted the use of buttresses, and there is evidence that the practice has been widely adopted. However, for the buttress to work it is essential that the bricks should be fully keyed into the main body of the wall. Unfortunately, some people are building the buttress as a simple column of bricks applied to the outside of the finished wall. This does nothing to increase the strength of the building, but the villager believes he is doing the right thing – 'because the engineer said so'.

In Yemen, another earthquake reconstruction project advocated the use of 'ring beams'. A ring beam is a continuous wooden or reinforced concrete beam which usually sits on top of the external walls to act as a collar to tie the whole building together. When an earthquake strikes, the ring beam enables all parts of the building to work together to provide mutual support. Again, this proved to be a widely adopted technology but in a significant number of instances the beam was only built on three of the four sides of the house. The Big Idea that the beam had to be a continuous loop had been missed while the house owners believed that they had success-fully implemented the idea since they had executed details which the intervener had presented as essential. By giving equal weight to *all* the details of how to execute the technology the 'why' of the idea had been lost.

The use of latrines is understandably perceived by the aid com-munity as an important contribution to improved health. More cor-rectly, latrines should be seen not merely as a potential benefit; they are also a potential threat. A sanitation technology which has been widely promoted over the last two decades is the Ventilated Improved Pit (VIP) latrine. As the name suggests, the pit under the latrine is ventilated by a chimney – this is the Big Idea. Less obvious are two essential practices:

- The top of the chimney should be capped with a fly-screen. The screen stops flies entering the latrine pit and, given that some flies will get in through the squat hole and lay eggs, it prevents flies getting out.
- The cubicle should be dark. If the cubicle is dark then any flies in the pit will head for the source of light at the top of the chimney where they will be trapped by the fly-screen and eventually die.

In the extensive literature on VIP latrines these two essential practices are stressed repeatedly. In practice, in many countries one encounters latrines which are ventilated by a chimney but which to a greater or lesser extent fail to observe the two essential practices. The fly-screen may be omitted or poorly fitted. Where it has been installed it may have rusted through or, if it is of plastic, decayed under the sun. The principle of the dark cubicle is widely ignored. Not only is the reason for it not immediately obvious but it also may seem to go against common sense; on the face of it, it would seem sensible to have a cubicle which is well ventilated and, hence, light and open in order to minimize smells. The result of the partially-appropriated VIP latrine may be a disease factory – a warm, protected breeding ground for flies which in some circumstances is arguably more detrimental to health than defecating in the fields.

In a sanitation project in Mozambique, the project designers appreciated the potential for failure inherent in the VIP latrine. They put their efforts of design, organization and promotion into a low-cost concrete squatting-slab for a latrine. Rather than ventilating the pit, their object was to seal it completely. The squat hole was provided with a lid which fitted tightly to keep smells in and flies out. The only essential practice involved was the need to place the lid back over the hole after use. The designers point out that whereas the VIP latrine was designed to extract smells through the chimney the use of the Mozambique latrine was smelly and so encouraged the user to replace the lid (Brandberg 1985: 25). The project was designed on the basis of the latrine slabs and their lids being made in the controlled conditions of large neighbourhood workshops rather than by individual householders. In this way, the essential features of the slab were dealt with in a controlled environment and so the opportunities for partial appropriation were minimized.

Simply because the potential for partial appropriation exists does not necessarily mean that a technology should be labelled bad. Rather, it means that if the potential benefits of the technology are sufficiently great to justify the risk then measures should be taken to predict and specifically counter tendencies to partial appropriation. The water closet has the potential for being the most hygienic type of sanitation available, yet it also has the potential for being one of the least. The Big Idea of the water closet is 'defecate in this'; the essential practice is 'flush it with water afterwards'. Countless examples may be found in which the Big Idea of this technology has been adopted but the facilities are not there to implement the essential practice.

SINGLE-MESSAGE TECHNOLOGIES

Other things being equal, the most successful technologies are liable to be those whose essence can be conveyed in a single message – there is neither a background deficit nor any essential practices. A technology being a single-message technology by no means precludes scope for a substantial number of messages regarding good practice. Good practice messages are those which may be completely ignored without jeopardizing the appropriation of the essence of the technology. Perhaps the pre-eminent example is the brick, a single message captures its essence: 'if you stack up a lot of these you get a wall'. The brick may be used in conjunction with a variety of mortars and laid in numerous ways with different degrees of skill, but the principle remains the same.

In Ecuador, one of the technologies promoted by Centro Sinchaguasin was a door which, instead of being hung from hinges, swung on pivots at the top and bottom. The reason for promoting this technology is that hinged doors exert large forces on the surrounding wall and this, if it is made of earth, is liable to break up. Before the introduction of the metal hinge from Europe, doors swung on hardwood pivots in stone or timber sockets. In Africa and Asia, as well as in Latin America, traditional earth wall construction can be seen cracking up around doors because of the adoption of the hinge without the adoption of the essential practice of more robust walling materials, such as concrete blocks. The failure of the earth walls is widely blamed on the inadequacy of earth rather than the inappropriateness of the hinge.

The promoted technology used metal bolts as pivots which revolved in holes drilled in the threshold and the lintel. The single message of the technology is 'you can hang a door on pivots'. The pivot door may be made of wood, metal or stone and may be rectangular, triangular, computer-controlled or secured with string. As with bricks, the conceptual core of the technology is robust since it consists entirely of the Big Idea. There may be limits to the applicability of the message; brick walls beyond a certain height and pivot doors beyond a certain width become problematic. The distinction is that such information comes within the bounds of common sense and trial-and-error experience. The knowledge that asbestos dust causes cancer cannot be expected to be reasoned from a villager's background knowledge and the Big Idea alone.

Building a reinforced concrete column may seem to require large amounts of information – the adoption of many messages. An engineering specification will lay down requirements of compressive strength, cleanliness of materials, the vibration of the concrete and the like. However, all these are good practices which can be almost totally ignored while still producing what the user-builder of a domestic-scale building will consider a satisfactory result. The construction of a normal column requires just one indivisible message: to make a reinforced concrete column, take four lengths of half-inch steel and tie them together at intervals with loops of steel and wire. Make a well-supported timber mould around the steel and pour in a mixture of cement, sand, gravel and water.

Clearly, this contains several messages but, unlike the messages of the VIP latrine, none of the messages can be independently applied in an apparently sensible way. All of the messages are contained in the Big Idea. The steel bars cannot be put together without the loops of steel or something similar; the steel structure has no function on its own, it only makes sense when contemplated in the context of the concrete; in the same way, the concrete column is useless without its reinforcement. Cases may be found in which concrete columns have been built without reinforcement but these are liable to fail, obviously and catastrophically. Again, to the hypothetical isolated community the message may not appear as a coherent sensible whole because it does not relate to a known result.

But, to large numbers of people, the concrete column Big Idea is an indivisible and self-sufficient package of ideas. The Big Idea provokes secondary messages concerning, for instance, the correct mix of cement, sand, gravel and water. However, there is a high degree of tolerance in these secondary messages which is what puts them into the category of good practice rather than background deficit. In the same way, the cement content of concrete blocks is highly variable yet the product remains recognizably a concrete block.

At least partial success in the absence of good practice is both the strength and weakness of reinforced concrete. In an earthquake or under abnormal loads, many reinforced concrete buildings in the Third World have collapsed which if built properly, in engineering terms, should not have done so. It may be that in a seismic zone a badly built reinforced concrete building is more dangerous than the previous traditional building. The dividing line between good and essential practice is not fixed. At the same time, it is apparent that in the eyes of the villagers their version of reinforced concrete is perceived as working. A series of procedures has been gone through which has resulted in a hard, tough, modern-looking, smooth, light grey column on which part of a roof or an upper floor may be supported. Their performance criteria have been met. The single-message nature of reinforced concrete has facilitated its adoption but resulted in vulnerable housing. A single-message technology is not necessarily a good technology, but it is a readily-adoptable technology.

Many technological ideas may be seen by their designers as indivisible Big Ideas while to the intended beneficiaries they are not. In Zambia, an improved block for building walls was introduced which, in order to function as the intervener intended, required the acceptance of two messages:

- Use a mixture of soil and cement.
- Use a block press to compress the block.

To the intervener, these two messages were inseparable since the desired result was a compressed soil-cement block. When the project was revisited some years after the intervention it was apparent that the first message had been adopted while the second one had not (Spence 1987: 51). People were making the blocks in traditional simple moulds but using the soil-cement mix. The resultant product was more resilient than the conventional earth

29

blocks, but less so than the compressed soil-cement blocks. The second message cannot be seen as essential practice since even when ignored an improved building technology still resulted. Similarly, if the first message was ignored and the second one acted upon an improved block would also result. There were two independently applicable Big Ideas.

If, from the outset, the technology had been treated as two technologies it is possible that a greater take-up may have resulted, while still leaving open the possibility for the full realization of the intervener's vision. One can envisage a situation in which soil-cement had been widely adopted in Zambia and a new project then launched to introduce compressed blocks as a further improvement.

Other cases exist in which elements of a technology are adopted and go on to produce a result quite different from that which the intervener intended. In Ladakh, in the northern Indian Himalayas, an NGO, the Ladakh Ecological Development Group (LEDeG) has been promoting a technology generally known as the Trombe Wall. Named after its French designer, the Trombe wall is a relatively simple means of heating a room using solar energy. A wall of glass, usually divided up into manageable panes with a timber frame, is placed a few centimetres in front of a thick masonry wall. The wall is painted black. The wall heats up in the sun and the heat is largely trapped behind the glass. During the evening the masonry wall acts as a storage radiator as it slowly releases heat into the room beyond. The Trombe wall technology would seem a sensible choice in Ladakh where the sunshine is intense while the winter temperatures are low.

Over a hundred of the Trombe walls have been built in the region but, to the best of my knowledge, all these benefited from financial assistance from LEDeG or other institutions. In other words, the technology has not been appropriated to the extent that people are prepared to invest their limited resources to bear the full cost of the technology. In the meantime, another technology has been rapidly and spontaneously adopted; this is known locally as the 'glass room'. Like the Trombe wall it involves a floor to ceiling wall of glass, but there is no heavy masonry wall behind it. In some cases the room is on a corner and both of the external walls are of glass, but in many instances just one wall is glazed. In the cold but sunny winter days, the room rapidly heats up allowing people to sit around and pass the time of day in relative comfort.

The history of the glass room is not altogether clear. It appears

Figure 2 A house with a glass room in Ladakh
The owners have added a glass room on the top left of the house. The floor
to ceiling wall of glass traps the heat of the sun and is a clear display of
status.

to be unique to the area so it is not an idea which has been imported
wholesale. Until a road was built in the mid-1970s, Ladakh was
largely cut off from the world. Fragile manufactured items, such as
panes of glass, were traditionally great luxuries which had to be
carried on pack-horses over high mountain passes. In the traditional
house of the more prosperous family, it is common to find glass
not in windows but reserved for the doors of a small display
cupboard enjoying pride of place where it can be seen by guests in
the main kitchen-cum-living room. When the road brought greater
prosperity and improved access to imported goods, it was not
surprising that an established status item such as glass became a high
priority. A few houses already had sitting rooms with floor to
ceiling shuttered openings. Progressively replacing wooden shutters
with glass windows would be an easy process, and for the majority
of households, which did not have such a room, there still existed
knowledge of the model of a light and open sitting place.
 Although the circumstances existed in which a glass room could

evolve, this is not sufficient to explain why it should have developed and spread so rapidly. Several villagers, when asked where the idea for a glass room had come from, said it had been introduced by LEDeG. The villagers explained how it was a way of using the heat of the sun. The workers of LEDeG deny that they were involved with the glass room's introduction.

The appearance of the glass room from the outside is much the same as a Trombe wall. It seems not unreasonable to believe that LEDeG's work in promoting the virtues of the Trombe wall has contributed to the rapid dissemination of glass rooms. There is an association in people's minds between four ideas: solar energy, walls of glass, LEDeG and progress. The Big Idea of the Trombe wall is, I suggest, being seen in two different ways by the aid interveners of LEDeG and the villagers. To the designer, the Trombe wall is first and foremost a heat-storing masonry wall to which is applied a glass skin to enhance the wall's ability to gather and retain heat. To the villager, the Big Idea is a prestigious wall of glass which traps the sun's heat. To put a wall behind the glass is not only a separate idea but, like the dark cubicle of the VIP latrine, it is an apparently nonsensical idea – why go to all the trouble and expense of building a prestigious glass wall and then put a solid wall behind it which keeps out the light and the heat and hides it from one's guests?

Through promoting the Trombe wall, not only has a different solution emerged but a different problem has been solved. The Trombe wall's advantage is that it keeps a room warm in the evening, whereas the glass room creates an additional habitable day-time room during the cold winter when much of the house may otherwise be uninhabitable. In addition, several villagers commented that when you go into a glass room on a sunny day it is really warm, you can bask in the heat, it is a luxury. A Trombe wall does not produce this kind of instant and gratifying result. The rapid spread of the glass room may have something to do with the promotion of the Trombe wall but, essentially, the glass room happened in spite of the Trombe wall promotion rather than because of it. An early consideration of the nature of the messages being promoted might have led to a realization of the benefits of the glass room and a change of strategy.

Other instances may be found in which the project workers have realized early on that the villagers were seeing the intervention in a different way to that intended and have modified the project constructively. Such a case is that of the Yatenga water conservation project in Burkina Faso (Oxfam 1987; Reij 1988: 74–7; Harrison 1987: 165–70). The Yatenga project, initiated by Oxfam, has been remarkably successful in rehabilitating barren, heavily eroded land. By placing lines of stones along the contours of the land, water run-off is slowed down, and so filtration is increased. Soil which would otherwise be washed away accumulates behind the stone lines and moisture lingers in the accumulated soil, triggering spontaneous growth of vegetation which helps to establish a more substantial stabilizing wall.

Behind the stone lines, soil and moisture continue to accumulate sufficiently for crops to be planted and sustained. The technique of stone lines is spreading rapidly, both spontaneously and with the encouragement of Oxfam's training staff. However, the stone line project grew from an initiative with a rather different emphasis. The Oxfam field director had been primarily concerned with planting trees and shrubs with a view to stabilizing the soil and providing much needed firewood. To this end he had introduced the agroforestry technique of 'micro-catchments' in which a small area of land, enclosed by a low earth wall, drains to a shallow basin in which trees are planted. The scheme worked well during the wet season but the trees still needed watering during the dry season so, due to the scarcity of water, many trees died. Several of the villagers, on their own initiative, chose to plant crops such as sorghum or rice rather than trees in the catchments. Since these were harvested each year there was no problem of plants having to survive over the dry season.

In response to these early experiences, the emphasis of the intervention switched from trees, the initial objective of the project, to methods of collecting and controlling rainwater. It then emerged that there was an indigenous technique of water harvesting using stone lines along the contours. This had fallen into disuse because of the difficulty of establishing precise contours. It is essential to follow the contours accurately since otherwise the water flows along the stone lines building up and causing damage until the line was eventually breached at the lowest point. Once this problem emerged, the project could then become sharply focused on developing and promoting a simple technique for determining the contours. The result was a 'water level' made from a length of transparent hosepipe attached to two stakes. What had started as a forestry project was transformed into one for introducing a simple surveying technique.

The Yatenga example demonstrates a constructive interaction between technical and material aid. The transparent hosepipe was not locally available, so the project bought tubing 'by the kilometre' in the capital which it distributed in lengths of a few metres in order to make the water levels (Oxfam 1987: 3). Hence, a highly cost-effective instance of material aid emerged from well targeted technical assistance. The project also demonstrates how, even in the indigenous process, a Big Idea like the stone line may emerge but fade away due to a background deficit, in this case the inability to establish reliable contours. By plugging the knowledge gap the project facilitated the development of a technology which was already latent in the community.

ESTABLISHING THE MESSAGES

The boundaries between the four categories of message – the background deficit, the Big Idea, essential practices, and good practices – may be fuzzy and shifting. Identifying the boundaries, or one's uncertainty about the boundaries, is liable to be a vital step in designing or evaluating a well-informed technological intervention.

A sanitation technology which has received a lot of attention from the Appropriate Technology movement is the composting latrine – a latrine which permits the user to recycle human waste as fertilizer. There are a number of different types of composting latrine. At Centro Sinchaguasin we speculated that aerobic composting latrines, which in other experimental projects have often suffered from too much moisture, could prove particularly suited to high altitude locations, such as the Andes, where the dry air encourages high rates of evaporation. In addition, at high altitudes there are no mosquitoes and relatively few other flies. The aerobic process, if it can be made to work, is more attractive than the anaerobic process, in which oxygen is excluded, for three reasons:

- Potentially, it is less smelly.
- The aerobic process generates higher temperatures which are important in killing parasites and other pathogens.
- The resulting moderately dry heap is easily contained in conventional above-ground constructions, while the wet slurry in anaerobic latrines requires either below-ground pits or above-ground tanks of concrete.

The latrine has two tanks which are used and emptied alternately to ensure that the faeces are fully composted by the time each tank is emptied. There are a large number of messages which need to be adopted for full appropriation of the technology, the principle ones of which are:

- Human faeces are a powerful form of organic fertilizer.
- Collecting and composting human waste and putting it on the fields at sowing time is more effective than defecating on the fields throughout the year.
- Adding animal dung to the human waste makes better fertilizer and makes better use of the animal dung.
- Adding vegetable waste results in better quality, less smelly and safer fertilizer.
- Using enclosed tanks will reduce health risks.
- Ventilating the tanks with a chimney improves the composting process and reduces smells.
- Alternating between two tanks improves composting and reduces health risks.
- The contents of the dormant tank should be turned regularly and kept reasonably moist in order to improve the quality of the compost.

The above messages are all of a technical nature and could be identified by a designer working at a desk or in an academic research station isolated from the community. They are products of the technology and display nothing other than the vaguest of notions of the characteristics of the target population. These messages do, however, give clear pointers toward a series of technical and social questions which need to be answered before the boundaries between the four categories of message may be determined.

For sociologists or anthropologists, all the above messages may be readily converted into questions of the form: 'is there evidence that this message already forms part of the background knowledge?' A second question is 'if this is part of the background knowledge is there any evidence that the villagers are not acting on this knowledge for some other reason and if so what?' For instance, in the case of the first message there must be few, if any, agricultural communities which are unaware of the fertilizing potential of human waste. It is, however, worth identifying and considering the message. If the target community is ignorant of the use of human faeces as fertilizer then this message becomes an important element in the background deficit. If, as is more probable, villagers are already defecating in their fields in a conscious attempt to raise soil fertility then it suggests that there is little point in making the use of human waste as fertilizer the principal message of a promotional campaign. In other situations, people may not be defecating in their fields because of social taboos rather than ignorance. In practice, sanitation programmes are generally run by urbanized professionals to whom the idea of human waste as fertilizer is novel and so this has become the focus of their promotion.

In the Hunza valley in northern Pakistan, there already exists the practice of collecting human faeces in simple latrines for use on the fields at sowing time. They have discovered that in the dry air and harsh sunlight of the mountains human and animal dung soon dries out and 'burns', and so loses much of its fertilizing potential. Soil fertility is a crucial issue for the Hunza people since arable land is scarce and the soil cover often thin. In neighbouring Baltistan and Ladakh, which are at even higher altitudes, the indigenous latrines are more sophisticated. Animal dung, leaves from trees and branches from bushes are put into the latrine. There is a high degree of awareness that the vegetable matter is an important factor in the quality of the fertilizer.

Sometimes, the sociological investigation of the messages feeds back into a re-framing of the technical messages. In some parts of Ecuador, careful farmers methodically collect the dung from chickens, rabbits and guinea pigs and store it in sacks until sowing time. The original messages envisage storing human waste and adding animal dung to improve the quality. In Ecuador, it may be useful to reverse the message and think in terms of the existing practice of storing animal dung and suggesting improvements through the addition of human waste.

The set of messages for a composting latrine define a solution. As suggested above, this is only one half of the designer's task. The second half is to assess the possibilities for, and consequences of, failure in the implementation of each message. What will result if a message is not acted upon? In the Ecuador composting latrine project all of the above messages were being promoted. As with the case of the Trombe wall, the latrines which were built in accordance with all of the messages were only achieved through the coercion of material aid. There were, however, examples of farmers building their own form of composting latrines. They had absorbed some of the messages and from these principles devised their own designs.

One man, whose farm was on a steep hillside of hardened clay, cut a chamber in the hillside with a squat platform above and an access hatch below which was bricked up until he wanted to empty the tank at sowing time. All of the family used the latrine, the animal dung went in as did vegetable waste from the harvest and the kitchen. The result was a well-managed single tank composting latrine which, in terms of its operation, was virtually identical to the indigenous latrines of Ladakh and Baltistan. When visited, the farmer was pleased with the compost and two of his neighbours were building their own copies of his design.

The farmer had adopted the Big Idea of collecting and recycling human waste and several of the associated practices, but had ignored the messages regarding ventilation and alternating twin tanks. The rejected messages either were not understood or they were discarded because the benefits did not seem to justify the additional expense and trouble. The question for the intervener is, was this a success or a failure for the project? In the context of high altitude Ecuador, are the practices of twin tanks and ventilation essential or merely good? The question for the micro-biologist and parasitologist is, does a poorly ventilated, single-tank composting latrine do more damage to the family's health than their prior practice of defecating in the field? If the use of such a latrine produces a situation which is worse than it was without the latrine then the project is a failure since it is doing actual harm. If it is not, it suggests that, given the initial state of the background knowledge, the project was being too ambitious in trying to promote the ventilated twin tanks.

THE PRINCIPLE OF THE BIG IDEA

The Big Idea of an intervention should be made explicit. Any indigenous deficit in the necessary background knowledge should be identified, as should any practices which are considered essential, in order to realize the benefit of the Big Idea. Bearing in mind that a development project which succeeds in introducing just one idea is doing better than most, the possibilities for, and consequences of, any of these secondary ideas not being adopted should be examined.

This principle requires the intervener to ask of every intervention, 'what am I really trying to achieve and how can it go wrong?' In order to answer these two questions it is generally necessary to ask a number of questions about the context. It is my impression that many of the useful technical questions surrounding Appropriate

Technologies do not yet have clear answers. For instance, the questions raised above, regarding composting latrines, have not been answered. The scientist can clearly describe a good composting latrine but he can not describe a bad one. The principle of the Big Idea demands that one establishes the conditions necessary for an idea to work rather than how the benefits of the idea may be maximized.

In an ideal world one would start from a complete model of the background knowledge of the target population. In practice, even with the most intensive anthropological studies, such a model is impossible to achieve. We do not even know what it is that we know ourselves. The best that can be achieved is a study of the directly relevant background knowledge, though it is only when a solution is identified that one can know what is or is not relevant. The validity of a solution can not be assessed without an understanding of the problem but, equally, the problem can not be fully investigated without some understanding of the nature of the solution.

Inevitably, the process of technology design and anthropological research is circular. The circle may be legitimately broken into at any point. The anthropologist may study a social problem because he or she has a hazy idea that technologies exist to cope with problems of that sort. Or a designer may conjure up speculative solutions to half-understood problems. It is common to mock 'solutions in search of problems', yet they have a place. Flexible plastic water pipe did not come about through detailed studies of water shortages, it emerged from fundamental scientific research into polymers. At whatever point in the cycle that the process is initiated, in order to progress the technology designer needs to make a first attempt at a design. Tentative answers need to be found to the following questions:

- What is the Big Idea?
- What information is it essential to have before the Big Idea may be reasonably implemented? What is likely to be the background deficit?
- What additional information must be acted upon before the new solution produces a situation which is better than that presently existing? What is the essential practice?
- What are some of the additional good practices?
- How might any of the above messages be misunderstood?
- How could the technology be misunderstood?

2

RECOGNIZED AUTHORITIES

'Why do you do this?' I would ask.
'Because it is good.'
'Why is it good?'
'Because the ancestors told us to.'
(Slyly) 'Why did the ancestors tell you to?'
'Because it is good.'

(Nigel Barley 1983: 82)

Who does what and what is done by whom? These are the questions with which each of us constructs a model of society. The social model has been described by sociologists as 'a web of competences' (Berger *et al.* 1974: 46). In stable societies, the locations and limits of skills and responsibilities are clear. One knows who to turn to with any given type of problem and what categories of action to expect from different people.

The web of competences describes a set of latent contracts between people. One party has a service which he can perform for somebody else given the appropriate incentive. When entering a contract we claim to be competent to deliver our end of the bargain, whether it be supplying goods, paying money or contributing labour to dig a communal ditch. In the context of aid, this suggests a model in which the intervener presents a clear service so that the villagers understand what is being offered by the intervener and what they have to do in return.

Changing societies bring changing responsibilities. In education and health care, school teachers and doctors have established positions for themselves within the social structure of developing countries. In other fields of development, the aid community has been largely unsuccessful in establishing new social actors whose roles are understood. The surly shopkeeper who sells asbestos-cement sheets and ironmongery with 'No Credit Given' is offering a more understandable service than the smiling development worker who wants to know your problems, be your friend, and raise your consciousness.

USING ESTABLISHED INSTITUTIONS

Peter Berger, the sociologist, suggests that 'the most fundamental function of institutions is probably to protect the individual from having to make too many choices' (Berger *et al.* 1974: 167). He goes on to suggest that the breakdown of many traditional institutions has led to 'the homeless mind' which he believes to be a characteristic problem of modern society. In the Third World, the speed with which institutions are crumbling is liable to be particularly disruptive to the individual's understanding of how society functions. Former power figures, such as religious leaders and local landowners, are being undermined while the arbiters and repositories of traditional wisdom and practice are being by-passed and discredited. Aid intervention has the potential to contribute to the disintegration of the understood web of competences as well as to reinforce and repair it.

Strengthening indigenous institutions is not an end in itself. Strong institutions are often the very obstacles to the just and equitable development which many believe the aid process ought to promote. Totalitarian states, powerful land owners, trade cartels, the caste system, apartheid and sexual inequality are all strong institutions whose further reinforcement may be undesirable. Many critics of aid base their criticism on the numerous examples where aid has strengthened the institutions of inequality and repression (for example Hayter 1981; Linear 1985). Notwithstanding this problem, the threat of the disintegration of the social fabric suggests that, where compatible with broader development objectives, it is preferable for aid projects to use and reinforce existing institutions.

The most obvious interaction between technological interventions and indigenous competences is in the matching of new technologies with existing artisanal skills. Problems arose in the construction of a village clinic in Pakistan. The design called for a timber ring beam located three-quarters of the way up the walls in order to increase the seismic resistance of the building. A stone mason or a brick-layer is used to the task of building timber lintels into a wall over door and window openings, this is an accepted part of his competence. The carpentry work necessary to make the ring beam was considerably more than that required to install lintels and the built-in ring beam was not considered by the carpenter to be within his sphere of competence. Although the work clearly required carpentry skills, the ring beam was neither a recognized task nor one to be undertaken at a conventional point in the building process. The carpenter was used to coming to a building after the mason and so building the roof structure and fittings around the finished walls. The construction process necessitated by the design for this building thus failed to mesh with the existing structure of skills. The design for the composting latrine in Ecuador, however, did mesh with existing competences. The design involved a built-in wooden seat which was pre-fabricated by a carpenter. This meant that the seat could simply be built into the walls in the same way that a mason would install a lintel. Both mason and carpenter had clear tasks which did not disrupt their understood work processes.

In village construction, the distinction between men's work and women's work may be clear to the villagers but obscure to the outsider. In Ecuador, different moments in the roof-thatching process are gender-specific. The women beat the bundles of cut grass in order to separate the stalks; they are also responsible for tying the wooden roof structure together with lengths of sisal. The men then use a large needle to tie the thatch to the structure with more sisal. When we came to promote a straw reinforced earth and cement floor mix, it soon became clear that the process of separating the straw stalks and cutting them to length was perceived as women's work. Even when male builders were being paid there was a resistance to performing such tasks.

The advantage of using existing skills lies not simply in the use of those skills but in the opportunity to exploit the complex network of existing and understood relationships within the rest of the community. Some improved cook-stove projects encourage householders to build their own stoves. Not only do these projects maximize the opportunities for technical failure but they require the intervener to establish a structure of dissemination which can convey detailed information to the individual householder. A stove project in Kenya took the approach of starting with an established existing product. The *jiko* is a small portable stove, made of steel, which is in widespread use in Kenya. The Kenya Renewable Energy

Development Project (KREDP) developed an improved *jiko* which was similar to the conventional one, but which had a ceramic liner. The new stoves could be made in the existing artisanal workshops and the ceramic liners provided new work for existing potteries. Thereafter, dissemination took place through the market-place and the informal structure of wholesalers and retailers (Harrison 1987: 210–16). By slotting into the existing system the improved *jiko* rapidly reached all corners of the country using the same mechanisms which bring Coca-Cola and aspirins to the remotest villages of the Third World.

The market is not the only route for technological change. In Nepal, UNICEF wanted to promote the use of Oral Rehydration Treatment (ORT). The technology was simple, the problem was one of dissemination. They identified two existing competences in the indigenous society related to health care. One was the *dhami*, the local faith-healer, and the other was *Durga*, a Goddess associated with cures. It was found that most *dhamis* when presented with a child with diarrhoea would prescribe withholding liquids in order to dry the child out. This is often fatal. Rather than trying to set up in competition with the *dhamis*, UNICEF chose to recognize the *dhamis* role in health care and support their activities while at the same time trying to persuade them to promote ORT instead of the practice of withholding fluids. To reinforce and complement this approach they produced a promotional card, the size of a playing card (see Figure 10, p. 99). On one side of the card were graphic instructions on how to mix the ORT, while on the other was a colour reproduction of a familiar image of the Goddess *Durga*. The use of the Goddess in this way served several functions:

- It associated in the villagers' minds the ORT with the competent Goddess of cures.
- It demonstrated to the *dhamis* that UNICEF were not acting in competition to their traditional methods and values.
- Since the *dhamis* were given stocks of the cards to distribute, their status was enhanced since the cards were modern, colourful and shiny.

In Ladakh, people regard certain Buddhist monks as the competent authorities to determine the location and orientation of the site for a new house. Each site is considered to have a particular spiritual animal lying beneath it. The monk has to determine what the animal is and how it is lying under the site. Once established, work has to start over the animals stomach. To an uninitiated architect or engineer the perceived competence of the monk in the building process may not be apparent. But, as in the example above from Nepal, one can envisage an intervention in which the role of the monks is exploited and extended by giving them instruction and appropriate educational materials to convey messages about well-drained sites and good foundations.

MATCHING TECHNOLOGY WITH COMPETENCE

For an idea to be adopted its characteristics must match the abilities of the intended user and of the intended promoting institution. Some new technologies and ideas which to an outsider appear attractive may fail to find an appropriate institution for promotion. Disseminating knowledge of nutrition is recognized by aid institutions as a major issue. Yet who should do it? Doctors dispense cures not recipes. Teachers teach reading and writing not cooking. And a nutritionist is not a person who has been heard of. Similarly, the solar cooker has been a recurring favourite of Appropriate Technology enthusiasts, but there appears to be no evidence of its wide-scale adoption. Some of its problems are certainly of a practical nature but it also has no obvious institutional home; who should make it and who should disseminate it?

The methane or 'bio-gas' digester has also been widely advocated. As in the composting latrine, human and animal waste is used to make fertilizer but, in addition, methane gas is produced and collected which can be used for cooking and lighting. After much promotional effort the experience from many countries has been that methane digesters can be a useful technology only at a certain scale of operation. In China, the larger, better organized communes successfully used digesters, while in India and Latin America some of the bigger farms are benefiting from their use. The individual poor rural household, at whom many programmes were initially aimed, was generally unable to successfully adopt the technology. The small-scale experimental household digesters have tended to clog up, produce little gas and in a few instances they have exploded. These problems were partly due to cost and practical considerations such as the lack of sufficient cow-dung to sustain the digestion process. However, as with the fibre-cement tiles in Chapter 1, the interveners were often expecting too much technological understanding from the village household.

Various Appropriate Technologies have failed because too high a level of quality control is required in a village- or household-level production process. This is not to suggest that village artisans are incapable of producing high quality products, as demonstrated by numerous examples, such as the manufacture of fine carpets, but that quality is dependent on a thorough understanding of the medium and the product. It is not sufficient to learn a new recipe, parrot fashion, during an intensive course. One needs a confident base of knowledge and experience from which problems can be solved and the nature of errors recognized.

One observer of cook-stove projects points out that zero-cost, 'build your own' programmes of stoves require each householder to become experts on combustion theory (Soedjarwo 1983: 232). He suggests that this is as futile as family planning promoters teaching people how to manufacture their own birth-control devices. Stove programmes, like that for the Kenyan *jiko*, which have concentrated the expertise in workshops and presented the consumer with a finished product have usually fared better. Each level of the operation has only needed to absorb the information it is capable of dealing with.

Where a technology already exists a network of relevant skills and roles will have coalesced around it. In Hunza, Baltistan, and Ladakh the composting latrine is established as an agricultural tool. The head of the household is the competent person to pronounce on how it should be used, what extra material should be put in, when it should be emptied and what crops could benefit most from the fertilizer. Other related competences have also developed. In Baltistan, children collect leaves to put in the latrine while women are responsible for carrying the fertilizer to the fields in baskets. In parts of Hunza, the task of emptying the latrine belongs to Baltistani casual labourers who come to the valley in search of work at the sowing season. In some places, it is reported that the emptying of the latrine would traditionally be an integral part of the spring-time religious festivals which coincide with the sowing. In this sense, the religious leaders have a role with respect to the timing of the emptying of the latrines. An educational programme aimed at the improved use of a composting latrine would have to recognize how any new knowledge could best plug into and exploit the established structure of skills, responsibilities and seasonal cycles.

Knowledge has to be matched with the appropriate competence

in two senses; not only does a new idea have to be targeted at the correct person but it also has to be introduced by an appropriate person or organization. If the type of knowledge does not coincide with the villager's perception of the nature of the source of the knowledge it may not be taken seriously. For instance, for the last couple of decades, the policy statements of health-care programmes for the Third World have repeatedly placed emphasis on prevention rather than cure. In practice, it has proved difficult to implement that policy. With the notable exception of vaccination programmes, the bulk of Third World health care remains curative.

The use of medical auxiliaries, community health workers and the like has been widely advocated (for example Werner 1977). The logical necessity for such paramedics is compelling yet the experience seems to suggest that, notwithstanding some notable exceptions, they have generally failed to live up to the early expectations. Nurses and midwifes have established a certain niche, but generally the epitome of the health worker is still that of the white-coated man who uses drugs to cure people and is called doctor; the villagers expect health care to be delivered by a real doctor in a real clinic (Hardiman 1986: 66). Any other image, role or name is drifting away from the definitive core – the Big Idea – of the understood competence. The alternative is not seen as 'appropriate' but merely as second-best.

Not only is it difficult for elements of a perceived competence to be transferred to somebody else, but it is also hard to assume competence in a field which is not seen as relating to one's currently perceived expertise. The doctor may know that the building of latrines is necessary for health but neither the villager nor, often, the doctor sees it as a task within the doctor's competence. The villager is unlikely to turn to the doctor for advice on how to build a latrine and the doctor is unlikely to be listened to if he gives that advice. Similarly, advice on nutrition and hygiene will often fall on deaf ears. The doctor will be listened to politely and respectfully but his advice will be ignored since it is perceived as being significantly outside his area of recognized wisdom.

The competence of the doctor overlaps and to some extent grows from that of the traditional healer. Despite the problems, possibilities do exist for establishing community health workers and, as in the case of the UNICEF project, for reinforcing and developing the traditional competences of local healers. In contrast, the agricultural promoter is a novel concept. There are people within a village who are respected for their skill and wisdom as farmers. Their experiences and experiments will be watched and emulated. The agricultural promoter frequently does not conform to this role model. The promoter is often a young graduate from agricultural college, with no land of his own in the village, who is actively trying to get people to change their ways. While you go to your doctor with problems the agricultural promoter tends to come to you for ill-defined reasons. For the agricultural promoter the problem is not simply how to do his or her job better, but how to establish in the eyes of the villagers a reason for existing. However, this very lack of an obvious role does create one advantage; since the agricultural promoter patently exists the villagers will try and ascribe a function to him or her. A person who is clearly being paid to do something but who has no function does not make sense. People will be curious and prepared to construct a sensible model of the promoter's function. The opportunity exists for the careful agricultural promoter to define a role from scratch.

The architect and the engineer are entirely products of the modern urban sector. Like the agricultural promoter, they have no direct equivalents in village life. The designer who is not also a builder most closely resembles the big landowner or his foreman who directs building works which are executed by the labourers and tenants. In recent years, with the increased phenomenon of migrant labour working on urban building sites, there are more people who are familiar both with the idea of architects and the nature of their role and their drawings. The villages are thus populated with some people to whom the word 'architect' means nothing and others to whom it means the person who produces the drawing of a modern house. In the same way that there are progressive doctors who are frustrated in their attempts to cast off their white coats and stethoscopes in order to become 'enablers' of community health, so too are there architects who, against their wishes, are locked into an immovable and predetermined role (see Turner and Fichter 1972: 122–47).

The architect produces the drawing and the modern builder, who has learnt to read drawings, builds the building. The recipe is simple and the competences understood. Between the elevated and distant architect and the local builder there is no other intermediary than the drawing. While the doctor can attempt to build on local healers, and the agricultural promoter can try to start from scratch, the 'barefoot architect', the building aid worker, the housing worker (or whatever he or she may be called, for as yet there is no established name) has to overcome this lack of comprehensible role at the village level.

UNDERSTANDABLE CONTEXTS

The contexts in which development education activities take place are often alien; the villager is put into social relationships which he or she cannot categorize under any recognized heading or where he or she cannot establish his or her own role. The classroom situation has been widely criticized for creating an 'us and them' relationship in which the teacher has knowledge and the pupils are empty vessels waiting to be filled. For villagers who have never been to school, the classroom format does not relate to any learning situation with which they have experience. At the same time, everyone is aware that schools are for children. By being placed in a childlike situation their competence as adults is under threat.

If a familiar social context can be created in which the villager can recognize familiar and acceptable roles both for him or herself and for the others present, an unfamiliar message or person is likely to be more manageable. This is a social equivalent to the ideas of 'background deficit' and the 'Big Idea' presented in Chapter 1. If there is no deficit in background knowledge, the villager will be equipped to test the validity of the new idea. In the Dhamar reconstruction project in Yemen, the teaching room was designed and fitted out to conform with the Arabic *mufraj*; the sitting room where men spend long afternoons chatting, smoking and chewing the local, mildly narcotic leaf. Carpets and mattresses on the floor created an instantly recognizable pattern in which any adult male felt comfortable in his role. Such a context was the natural one in which novel ideas were considered at leisure and problems could be shared and dissected with one's peers.

In many places the offer and acceptance of food and drink plays an important part in social intercourse. When one visits a villager's house it is normal to be offered some kind of refreshment, from a Coca-Cola to a substantial meal. However much one may explain that one is not thirsty or that one has just had lunch, the food will still be prepared. The offer and the acceptance of food establish the roles of host and guest. Both anthropologists and development workers recognize that turning down such offers of food and drink is liable to give offence and damage the nature of the relationship.

Despite this commonplace recognition, in situations in which the roles are reversed the significance of courtesy refreshment has been largely neglected. If a development institution is running a course, provision will normally be made for meals for the participants as a matter of simple functional need. But when a villager visits a development institution's office or a technology demonstration centre for a short visit, it is unusual for food to be offered since there is no compelling physical need.

At Centro Sinchagusin, we realized that it was important for us to offer refreshments to our visitors so that they could establish their status as guests and so that the Centre's domesticity could be expressed. The provision of food is not in itself novel. The important element was the use of food to reinforce a situation which was, in image and in practice, domestic. The domesticity was further underlined by the presence of children and domestic animals. The social context of being a guest in the home of someone who had a few rather interesting and unusual examples of domestic technology was understandable; being a visitor at a demonstration centre was not. The identical technologies could have been displayed in other circumstances and could have been demonstrated by friendly people, but if the context had not been that of someone's home it would have been significantly more alien. Although the project was called Centro Sinchaguasin and the name appeared on letterheads, bank accounts, project proposals and publications there was no sign-board at the Centre which declared it to be a Centre. Nor was there a fence or a visitors book. Villagers were invited to visit 'our house', not to visit Centro Sinchaguasin.

Occasionally, other development institutions would bring large groups of villagers to visit the Centre. Inevitably, such visits had an air of artificiality. It could not be the same as one householder dropping in on another. Yet, within this broadly alien context, it was still possible to establish understandable social handles for the visitors to grasp. The visits were timed to start late in the morning and end in the middle of the afternoon. On arrival, the visitors would be given a hot drink and a sweet roll. There would then be a formal tour of the Centre which was not a particularly comfortable social experience since, by its nature, it was unnatural. There tended to be few questions and a tangible nervousness. The formal tour was not intended to serve as a comprehensive education, but as a first exposure to the technologies on display.

The tour was followed by a full lunch which became the pivotal point of the day. The meal firmly established the visitors' status as guests and provided an understandable social context for idle chat. Once social confidence was established, conversation turned more readily to the technologies. After lunch, people were free to wander about and there would be members of the Centre's staff, with whom they had eaten, available for questions, including at least one woman. The afternoon session became the principle learning period. Women visitors who were invariably silent in the formal tour of the morning would start revealing which problems and solutions they were interested in. They would express opinions and seek out details. At the end of the visit there would be a final hot drink and sweet roll during a last group discussion.

Due, perhaps, to the unhappy history of insensitive formal education there is currently a great emphasis in development education on the use of informality, participatory discussions and humour. Laudable though this is, a situation of friendly socializing is not the only context in which people both understand their position and are able to learn. A readily understood competence is that of the client with respect to the tradesman. A client is someone who, within understood limits, is competent to instruct one to perform a task. Often a tradesman will disagree with a client, or think that the idea is daft, and may well say so. But if the client insists and if the task is not in some way demeaning, the tradesman will execute the task. In so doing, the tradesman will be exposed in detail to a new idea which he is free to reject but which he may also adopt. The situation is a controlled one. Client and tradesman do not even have to like each other but, unlike the classroom situation, it is an adult relationship. The client is employing the tradesman because he respects and recognizes his competence. The tradesman knows why he is there, knows that the client is paying for a specific service and is comfortable in his competence to perform his job, with the one possible exception of the unusual task. The new task becomes a challenge within the context of his competence.

In everyday life, many new skills are learnt on the job. Unlike the classroom or the currently fashionable 'workshop', it is the normal way of acquiring technical skills. The possibilities for development projects using employment as a dissemination strategy have been largely ignored. The language of development is all about participation; employment smacks of the unsavoury exercise of power. Yet, the power of employment is only being used to expose the individual to a new technology, the decision about whether to adopt it is entirely his or her own. At Centro Sinchaguasin we employed a builder to lay a straw-reinforced earth and cement floor. At the time he was sceptical. Later, he was observed bringing friends and potential clients back to show them the floor which he had made, the construction of which he explained in a proprietorial way.

In a project to build a rural teacher training centre, an Ecuadorian NGO, FUNHABIT – *Fundación Ecuatoriana del Habitat*, decided to build the rammed-earth walls using moulds which consisted of a set of wooden panels which could be bolted together in a variety of ways in order to make a variety of 'L' and 'T' shaped corner moulds as well as straight moulds. In all, the designers envisaged five different configurations. Because it was a large building they had four sets of moulds made and they employed several builders. The builders, once instructed to build using the new moulds, were in the understood context of having a job to do and certain tools available with which to do it. On their own initiative, the builders put the four sets together and during the course of the building work identified thirty-four configurations for the moulds. Rather than passively accepting instruction on the use of the moulds they had taken it as part of the job and made it their own.

NEW ROLES

A solution where an aid intervention exploits and reinforces an existing competence may not always be desirable. One example might be a credit programme. In most, if not all, societies, the idea of credit is familiar. Versions of the social institution of the money-lender or the pawnbroker are found across the world. A socially orientated rural credit programme may well find it unacceptable to channel its funds through the existing institution. In other instances, the nature of the intervention may be so novel that it does not conform with the understood characteristics of any existing competence within the community. Perhaps the pre-eminent example is the introduction of electricity. The nature of an electricity supply demands that a specialist organization is established to disseminate, maintain and sell it. Regardless of any procedural incompetence in the execution of the task of supplying electricity, the electricity companies of the world have rapidly created a location for themselves in the web of competences in the minds of hundreds of millions of people. The purpose of the organization is clear to all concerned. Few organizations can attain such clarity.

Many development organizations will have to take on numerous tasks, and as the number increases so institutional clarity will tend to decrease. The purpose of the more general rural development organization is often obscure even to its own senior staff. To the villager and the isolated field worker, it may be a complete mystery. The motive for establishing a rural development organization may be a clear and legitimate sense of outrage at gross injustice and poverty. Once established, the staff of such institutions may continue to share the bewilderment which most of us feel most of the time regarding what actually to do about the overwhelming problems. From the villager's perspective, where he can see that an organization exists, in order to make sense of the world he will try and ascribe a competence to it. In Ladakh, LEDeG appears to have become associated, albeit vaguely, with competence in 'things made of glass'. Even though it did not introduce the glass room this achievement has been widely attributed to LEDeG since people have tried to construct a model which makes sense of LEDeG within their model of society.

In Ecuador, an NGO, the *Centro Andino de Accion Popular* (CAAP), implemented a community-based, post-earthquake reconstruction project in eighty villages. CAAP is based in an old hacienda building which had been abandoned shortly after land reforms a decade earlier. Several months after the earthquake the communities chose to express their gratitude to CAAP by holding their festivities for the big religious festival of the summer in the grounds of CAAP's headquarters. In the years since then they continued to celebrate the summer festival at CAAP. It emerged that this had been a tradition in the times of the hacienda whose building CAAP now occupies. The communities, in trying to ascribe an understandable competence to the development institution, had, for better or worse, taken the hacienda as the closest existing category.

A study by the development writer Paul Harrison concluded that successful projects are frequently to be found where 'a single donor dominates the field concerned, and has single-mindedly pursued the same objective without wavering' (Harrison 1987: 314). He cites as examples two soil conservation programmes in which foreign donors have provided nationwide support over the long term in one very narrow clearly defined technological field. Anybody in the country, from the high official to the villager, would know who to turn to for technical help in soil conservation measures. Equally, anybody meeting representatives from the organizations or seeing their vehicles would know what it is that those people do. They have a place in the order of things.

The Mozambique latrine project, mentioned in Chapter 1, created a similarly clear competence, not at a national level, but across the city of Maputo. By focusing on the production and dissemination of one tangible entity, the latrine slab, not only was the technology clear but so too was the competence of the promoting organization: 'they're the people that sell latrine slabs'. A clear competence does not necessarily have to relate to everybody since it may not be relevant to everybody. The promotional strategy for the Kenyan improved cook-stove, the *jiko*, as we have seen, exploited the competences of the conventional market, but the Kenya Renewable Energy Development Project (KREDP) also established a clear role for itself by offering training to the metalworkers and ceramic workshops.

The single-minded approach carries with it the danger of barrel vision. It is all very well to pursue one objective without wavering so long as it is the right one. The above examples were all preceded by exhaustive periods of product development and field trials. This implies that the desired competence will be different depending on which stage of the process is being executed. The implementation requires a clear-cut service, while research and development needs to positively avoid being perceived as promoting specific solutions since, if prototypes fail, they are liable to undermine the credibility of any subsequent implementation programme. Research and development is not an activity which marries well with the idea of a society composed of a stable network of competences. To some extent, someone researching a problem is someone who does not have the solution.

At Centro Sinchaguasin, people were at times mystified by what we did. Who were we? What was our purpose? In retrospect, it seems that the housing work of the Centre was saved by its association with a forestry project. From the outset we had considered there were benefits to be had in running a forestry programme alongside a housing improvement programme; visitors who came to see the housing experiments would be introduced to the nursery of native tree seedlings while visitors that came for trees would see the buildings. These benefits certainly occurred but, in addition, the tree nursery which was a straightforward implementation programme established an institutional reason for existing which could be readily understood; 'those people produce and sell native trees'. The housing technologies were seen as experiments which forestry workers were carrying out on their own homes.

In recent years, agricultural development workers have been pointing out that village farmers do not just blindly follow the practices of their ancestors. Research and development is going on all the time. Individual farmers try out new seeds, new fertilizers or new combinations of plants in a systematic and scientific fashion when the real work of the day is over (see, for instance, Chambers *et al.* 1989). Individuals will try something out in their own home and they and their neighbours will patiently wait to observe results. The experiments of the man who built a single-tank composting latrine in Ecuador were watched by his neighbours. The following year, once success had been demonstrated, the neighbours built their own. The home is the naturally understood context for research and development. The full-time scientist and researcher is a luxury of urbanized society. The implication is that an organization which wishes to carry out research and development would be best advised to carry it out either out of sight behind closed doors on a university campus, or as an appendage to a 'real' understandable job in the field. The field-based, full-time researcher may feel more in contact with the research and development, but may be incomprehensible to the villagers.

Within the slowly evolved network of competences in traditional society there are those which relate to conveying knowledge and wisdom rather than to immediate material gain. But a newly-created competence is likely to need the benefit of a tangible product. It is easier to create a competence related to material aid than to technical aid. Formal education often revolves around the gaining of certificates which are needed for jobs rather than the value placed on the knowledge and wisdom acquired in a school. A competence for providing a physical product may be used as a hook around which other competences in the provision of knowledge may be hung.

One of the clearest examples of such a strategy is that of the Aga Khan Rural Support Programme (AKRSP) in northern Pakistan. The villagers will be asked as a group to suggest a single infrastructure project which, if realized, would economically benefit the bulk of the population of the village. Typical of such infrastructure projects would be a bridge, a major irrigation channel, a vehicle track to link the village to a main road or protective works to prevent valuable land being washed away in the annual flooding of the rivers. Engineers and other specialists from AKRSP will assess the viability of the project and prepare a budget. Once approved, AKRSP will pay for material inputs and also pay for labour. However, the wages will be paid into a communal bank account in the village's name rather than to individual villagers. Thereafter, not only will the villagers be free to use their communal capital as collateral for loans for other projects but they will have been exposed to the technical expertise within AKRSP and encouraged to use that resource to develop their own projects.

The service being offered by AKRSP is not as clear-cut as that of the electricity company, but it is still clear. Rather than presenting themselves as a development institution there simply to 'listen to the people', they have presented a clear offer which if accepted gives the villagers rights and responsibilities. It is a contract in which each side knows what it has to do and what it expects in return. This approach by no means precludes listening to the people and providing a range of additional technical assistance, but the institutions' role as a partner in a clear contract is established.

THE PRINCIPLE OF RECOGNIZED AUTHORITIES

Each person has ideas about who does what and what is done by whom. Often people do not understand what development institutions do. Try and offer a clear service and where possible introduce technologies through people, organizations, and situations whose purpose and authority is recognized.

The Principle of Recognized Authorities requires that, both before and after a proposed intervention, the intervener should ask: 'in the eyes of the villager, who does what and what is done by whom'. It implies that in order for an intervention to be successful it is useful for the intervener to do three things:

- Identify the existing network of social institutions in the village society which may be relevant to a planned intervention.
- Relate one's intervention to individuals, organizations and situations whose purpose, skills and responsibilities are clearly understood by the villagers.
- If it is absolutely necessary to introduce a new actor or a new situation into village life, ensure that its purpose is made as unambiguous as possible.

3

MAXIMUM SERENDIPITY

My object was, not to see inns at turnpike-roads, but to see
the country; to see the farmers at home, and to see the
labourers in the fields; and to do this you must go either on
foot or on horseback.

(William Cobbett 1830)

If the object is to present intended beneficiaries with 'reasonable'
technical options then the intervener has to understand the framework of reasoning of the target population. An important element
of any technical assistance programme is likely to be research into
the nature of the problem as perceived by the intended beneficiaries.
However, the important information is likely to be that which one
does not know that one does not know. Research based on
questionnaires and meticulously-designed programmes of investigation presuppose that one knows what questions to ask.

OBSTACLES TO RESEARCH

The problem and strength of conventional anthropological and
rural development research has been in its emphasis on in-depth
examination of individual locations. The importance which has been
placed on such research can in part be seen as a reaction to the
discredited notions of universally applicable technical solutions
which were implicit in some earlier development aid. Such solutions
are often considered to have failed due to a lack of understanding
or consideration of the social, cultural and economic context. More
recently, there has been a reluctant reaction against the later,
highly specific studies and the realization that, notwithstanding the
necessity for cultural sensitivity, there is a need to act on a large

scale, and hence a need to gather information on a wide front in a short period of time with limited resources.

The necessity for an alternative research approach was identified by Robert Chambers, Gordon Conway and others in the late 1970s and early 1980s. In his book, *Rural Development: Putting the Last First* (1983), Chambers described how research procedures and institutional structures were leading to a systematic under-perception and mis-perception of rural poverty. Chambers' critique can, I believe, be encapsulated in two statements:

- Many potentially useful observations and insights are ignored because they do not conform to what are considered respectable standards of data.
- Much rural poverty and the reasons for that poverty go unobserved because of the built-in biases of 'rural development tourism'.

Rural development tourism is the term which Chambers uses to describe the manner in which the urban-based researchers and holders of power conduct their brief visits to rural areas. Chambers identifies six biases which influence how rural poverty and development are observed and described by visitors (Chambers 1983: 13–22):

1 **Spatial biases: urban, tarmac and roadside** The concern for the comfort of the visitor, the road system, the time and fuel available, the location of 'suitable' overnight accommodation and the urban centres of reference all conspire to focus attention on communities which are close to the better roads leading from the major cities.

2 **Project bias** Visitors are guided by the information and the contacts available. These generally relate to already-existing projects. As more people visit these projects, write about them, lobby for them and weld them into the international network of contacts, a self-reinforcing mythology is established. Attention and funds become ever more focused on an increasingly-atypical pet community. This bias is readily discernible from the literature, in which the same handful of projects keep reappearing.

3 **Person biases** Key informants are the most articulate and are thus atypical. Such people are, like the visitors themselves, generally male. They are village headmen, religious leaders, teachers and the like, whose views are often recorded as representing the community opinion. Informants are also available to inform – they are not dying, working long hours in somebody

else's workshop or temporarily working in the city. Visitors to an area are often accompanied by one of a handful of project workers who speak the visitors' language. Their views become a filter for outside perceptions of the project.

4 **Dry season biases**   Problems of disrupted transport, discomfort, lack of time and inflexibility of programme conspire to influence the visitor to schedule his trips to fall during the season of most clement weather. Hence, many of the most severe problems facing rural people go unobserved since they are seasonal.

5 **Diplomatic biases** Politeness, timidity and cowardice can prevent the real problems being exposed and confronted. Many visitors are in some sense senior people being guided by junior people. To seek out and examine the worst problems and particularly the failures of aid programmes is often seen by both parties as well as third parties as a process of criticism of the junior person.

Among Government organizations, NGOs and anthropologists alike, there are field workers who are intensely proprietorial toward *their* people. Diplomatic expediency can effectively ensure that no visit is made to a community unless chaperoned by its 'champion'.

6 **Professional biases**   The specialization of interest encouraged in professional training and attitudes has a strong tendency to create a tunnel vision which discourages an understanding of the links between issues. It also leads to a predisposition to regard one's own thematic area as the central issue.

The standard response to this kind of criticism is that we should 'listen to "the people"'. This practice can be hard to implement, particularly for the visitor. Even when one's professional bias is not acting as a filter to what is heard, there are many other obstacles to successful listening:

- Language
- Suspicion
- The villager's pride
- Unspoken assumptions of villager and researcher
- The villager's desire to please
- The villager's expectations
- The perceived role of one's companions
- Limited time
- The dominance of the articulate.

Language

For most of us, to listen to a conversation in anything other than our first language will mean, at best, missing the subtle nuances. In Northern Pakistan there are five principal languages as well as many dialects and the notional *lingua franca* of *Urdu*. Even for Pakistani workers and researchers there is frequently a need for a translator. In some instances, for a non-*Urdu* speaker, comments may have to pass through two translators. In such circumstances, any notion of the visitor casually listening in on a free exchange of views must be illusory. In *Quechua*-speaking communities of Ecuador there are both problems and benefits to be derived from language difficulties. Most men in the communities speak a form of Spanish. Many women only speak *Quechua* and so are further marginalized in the negotiations with the Spanish-speaking workers of development institutions. On the other hand, to a primarily English-speaking researcher there is the advantage that the villagers are themselves speaking in Spanish as a second language and hence they have an understanding of the linguistic problems. Conversations take place in a simple and slow Spanish for the benefit of both sides. Indeed, for the middle-class, city-dwelling Ecuadorian aid workers, their relatively sophisticated Spanish may be a handicap if not kept in check.

Suspicion

The visiting aid worker may be a focus of suspicion, and the researcher even more so. If the villagers are uncertain as to why the visitor is there it is unlikely that the visitor will be in a position to listen to and benefit from genuine expressions of needs and concerns. It is rarely adequate simply to explain one's presence, since the explanation is itself likely to require concepts which are outside the villagers' model of how the world is structured.

In Ecuador, it was found that the foreign aid worker can sometimes have an advantage over his professional Ecuadorian counterparts. Sociologists have pointed out that the stranger or the foreigner may be more approachable for 'from the point of view of the approached group, he is a man without a history' (Schutz 1964: 97). The bulk of Ecuadorian workers are *mestizos* (mixed race) or *blancos* (pure race descendants of Spaniards) who are burdened with all the historical associations of the large feudal farms and of the bureaucracies and injustices of the modern state. However, as the villagers acquire more experience of foreigners and suffer more disappointments from unfulfilled promises, this potential advantage is being dissipated.

Pride

Generally, the visitor is interested in identifying problems. It is only human to feel that to admit to an outsider that one has problems involves loss of face and hurt pride. This may lead either to a denial of any problems or a concentration on problems which, for various reasons, are considered less shameful. Particularly shameful problems may relate to issues such as sanitation or income. In Ladakh, Zanskar and Baltistan, traditional practices of sanitation were found to be, as yet, largely free of shame. In Hunza, which has been more exposed to modern culture and has less sophisticated traditional sanitation systems, shame led to a reluctance to discuss or admit to the practices which probably constitute the principle housing-related problems in the district.

Unspoken assumptions

Visitor, local worker and villager will each bring to a conversation different sets of unspoken assumptions, sometimes described as tacit knowledge or mind sets. The false assumptions brought by visitors have been widely discussed elsewhere (for example Gupta 1989: 24–31). An example is the commonly held and frequently false assumption that the principal farmer of the household is a man. Less well understood and accommodated are the assumptions which villagers have about the visitor and about the aid institutions they encounter. The literally extraordinary level of ignorance which the honoured visitor may have of day-to-day issues, and of the motivations behind people's actions, may be beyond the comprehension of the villager. Crucial information may not be given because it is considered too obvious. Equally, the visitor is not likely to ask a crucial question if he already believes the situation is clear.

In northern Pakistan, UNICEF have built a large number of drinking-water installations in which clean spring water is piped to village stand-pipes. This is considered by the aid establishment to be a major contribution to tackling the health problems associated with using water which has been taken from open and polluted irrigation ditches. In one village, a stand-pipe stood next to an irrigation ditch and a group of villagers were asked the question, 'from where do you get your drinking water?' The question appeared stupid to both parties since the answer seemed obvious. The answer was from the irrigation channel. It emerged that running water is considered clean, and cold water is considered particularly clean. Since the irrigation channel is fed from glacial melt-water it is cold, while the spring water is relatively warm and thus dirty. To the villager, the answer was thus obvious and rational yet the contrary answer seemed equally obvious and logical to the outsider with a different rationale. Studies in other parts of Pakistan have revealed that this was not an isolated instance.

The sets of assumptions which locally-recruited aid workers bring to bear on a situation have also been neglected. Since these workers are often acting as guides and translators they can exert a large influence on research through the selection of informants and locations which they believe, perhaps falsely, conform to the needs of the visitor. They may also be over-confident in that, being from the area, they already know it all. In a mountainous region such as northern Pakistan, the multitude of micro-climates are mirrored by micro-cultures which may vary from valley to valley and from village to village. The assumptions of a worker recruited from one village may unconsciously conspire to conceal significant differences found in another village. One example came from a highly experienced local field worker who, in a neighbouring valley, encountered the phenomenon of cattle sharing the main room of the house with the people during winter, a practice which he firmly believed had died out a generation before.

Desire to please

Villager's assumptions of the visitor's motivations are often reflected in a polite desire to please. This may manifest itself in attempts to give the right answer to the question or to what they see as the implicit question. Those topics of conversations may be chosen which the villagers believe the visitor is interested in. The visitor does not have to be guilty of dictating to them or asking leading questions to influence the flow of information. His presence, appearance and perceived associations can be enough.

In the context of an ongoing relationship with an aid institution, villagers may express enthusiasm for what they perceive as the field worker's pet scheme of the moment with a view to increasing the chances of access to greater benefits in the future. In Pakistan, there are examples of appropriate technology interventions in agriculture and sanitation in which the apparent take-up appears to be a case of the villagers wishing to please the aid workers in order to be classified as a progressive and active village. In Ecuador, where many of the indigenous communities have been exposed to a wide variety of different aid institutions, some of the village leaders have become highly skilled in expressing their problems in a terminology which they know will appeal to the donors.

Expectations

For the villagers, the fact of an outsider visiting and wanting to know about problems is generally a clear indication that the outsider, or somebody connected to him or her, is offering to give material help to do something about those problems. Not unreasonably, the villagers will consider this to be the only logical explanation for such questions. This not only creates potentially damaging expectations of assistance which may not come, but it is also likely to steer the conversation into an aid-orientated mode, with a concentration on issues which are male and material.

One's companions

For visitors, a one-to-one conversation in the village is relatively rare. Even if he or she speaks the local language, normal social practices combined with a natural curiosity towards the visitor will make group discussion the norm. What an individual villager says is liable to be coloured both by his perception of his own companions and of those of the visitor. A poor member of the community may be reticent about his problems in front of his wealthier neighbours, or before the village leader. If, as is likely, one or more of the visitor's companions are aid workers already known to the village, the conversation may be designed to please the aid worker. In Pakistan, some houses were visited with housing improvement workers while others were visited with health or agricultural workers. It was found that when a housing improvement worker was present the recipients of improved cook-stoves and latrines generally reported satisfaction with these new technologies, but that when a housing worker was absent the bulk of the reports were more negative.

It was consistently found that when I accompanied a health worker on a visit to the house of a patient the subsequent interview progressed in a particularly relaxed and informal way. I, as the researcher, was not the focus of attention; the health worker and the patient were. The patient's problems provided a natural opening to the conversation and my presence was readily categorized as that of companion to the health worker whose presence was well understood. With other people, such as housing workers, agricultural promoters and social organizers, our presence could not be so easily explained.

Time

Chambers advocates not just listening but also sitting. He uses this term to emphasize the need to make time to relax, break down barriers and allow issues to slowly emerge. A special approach is that of 'total immersion' used by anthropologists who live in a single community for a period of years. Yet, experiences show that even such lengthy periods of detailed study may later be found to have missed key features of a society (see Turnbull 1984: 125–9). For the most part, the research of the outsider will remain a relatively hurried business with all the artificiality of a special event.

The articulate

Perhaps the greatest obstacles to successful communication are the articulate: the head man or school teacher who insists on attending every interview and 'explaining' his fellows' views; or the 'man of the world' who knows how outsiders should be entertained and, out of hospitality, will not let the guest be bothered by the 'ignorant womenfolk'; or the progressive farmer who knows how to make links with development institutions and is interested in channeling material aid to *his* farm. Often the only way to break through this barrier is by causing offence, which in many circumstances can be too high a price. Similar problems may be encountered with guides and translators who choose to answer one's questions directly rather than put them to the villagers.

ENCOURAGING CHANCE LEARNING

To try and overcome the above kind of problems, a number of techniques have been devised which are now frequently grouped under the title Rapid Rural Appraisal (RRA). The results from using techniques such as workshops, semi-structured interviews and chain interviews can be good, bad or indifferent depending on luck and the skills and attitudes of the practitioners. Whatever the method chosen for conducting a village visit, it is probable that, owing to the nature of an outsider's visit to an isolated community, the predominant technique actually employed will be the *de facto* semi-structured group interview.

The fast-accumulating case studies of RRA suggest that there will be an increased likelihood of good-quality material emerging if, rather than thinking in terms of specific events, one is constantly trying to expose oneself to situations where chance learning is more likely to occur. The proposition is that some situations are more conducive to chance learning than others. As Chambers puts it, 'It is by talking, travelling, asking, listening, observing, and doing things together that [aid workers] can most effectively learn from one another' (Chambers 1983). Rural researchers cannot predict what, when and where they will learn something, but they can reasonably suppose that some situations expose them to more potential learning situations than others. Chambers illustrates his point by an example from Africa in which a researcher travelling to a research site learned the key information from his companions in the vehicle on the way. The chat in the back of a Land Cruiser does not fit well with rigid research protocols and learned papers, but it may be more useful.

Anthropologists have long advocated the use of participatory research, which is generally presented as a method for investigating a particular task. The anthropologist helps with the potato harvest in order to find out about the potato harvest. In practice, it is a lot more than this. It is an excuse for spending time alongside other people and maximizing the opportunities for the unexpected. In general, opportunities for fruitful informal interviews may be maximized by engineering situations in such a way that they provide the following two characteristics:

- **A tangible focus** In a conversation between a villager and an interviewer it is not unnatural for the interviewee to feel as though he is under the microscope. The inevitable tension which results can be partly defused by shifting the focus to some tangible object. The focus may be an educational aid like a poster or a flip-chart. There are a number of graphic manuals on the use of graphic materials in workshops and the like. But, wherever it is possible, the real thing is usually the best communication tool. A household interview may start simply by talking about the household objects around or the house itself. A visit with a health worker has an obvious focus presented by whatever is ailing the patient.

- **Creating understandable contexts**   In the previous chapter it was suggested that understandable contexts were important for absorbing new ideas. The same applies to divulging information. Many interviews are carried out in circumstances which are alien to the villagers. The surroundings may be their own house, but the experience of being interviewed or participating in a workshop does not relate to their day-to-day experience. Similarly, the relationship with the interviewer or workshop 'animator' does not correspond with any understood model of a social relationship.

A British architect, John Sanday, while working on conservation work in Kathmandu tried to find out how the traditional fine tapered bricks used in the old temples were made. Using conventional sources and simply asking around he had no luck. He set up a brick workshop *ad hoc* in the middle of a town square using his best guess at the technique. A crowd was soon attracted which included several old men who proceeded to debate furiously as to the correct way of doing it. Eventually, through discussion and experiment a consensus and a workable technique was established. The physical focus on a practical task and the context of market-square debate was something which everyone understood.

Once an activity and a relationship have been categorized in a familiar and secure pigeon-hole, more adventurous explorations of one's circumstances may be carried out in safety. The junior field worker travelling in the back of a Jeep with an esteemed visitor is in an understandable situation, travelling from A to B. He realizes that his fellow traveller, like himself, has nothing to do and so the worker feels free to converse informally. Outside the vehicle, at a project site or in an office, he is once again a junior worker to be seen and not heard unless asked a specific question. Casual encounters sitting in vehicles and walking along paths allow for a free, natural and unhurried exchange of information and opinions which rarely occur sitting around a conference table.

Working together on a shared task is an understandable context. The community infrastructure projects of the Aga Khan Rural

Support Programme in Pakistan are largely successful because of their simplicity. They do not require a great deal of research before they are executed. But once they are being carried out they are in themselves an excellent context for casual research. The day-to-day process of moving materials and equipment and arranging the logistics of the operation generates a wealth of opportunities for the field staff to better understand the problems and priorities of the villagers.

Walking or wandering around is possibly the pre-eminent development research technique. Its banality belies its value and explains its lack of prestige as a research methodology. Few field workers would deny its value but under the pressure of 'real work' it is an activity which is often neglected, and where it happens there is no format in which the product may be respectably displayed. To remedy this, a number of practitioners have devised formalized walking practices.

One of the most striking examples is the 'group trek' used by the Agricultural Research and Production Project in Nepal (Mathema and Galt 1989: 68–73). This is a Government project supported by USAID. In a typical group trek, a multi-disciplinary team of half a dozen people spends three days walking across a target area carrying out group and individual interviews. Interviews always include one with the village headman, who will be asked to identify key farmers for subsequent interviews. Each evening is dedicated to a group review of the day's findings and the identification of interesting lines of further enquiry as well as gaps in the information. Both for discussions within the team and with villagers encountered on the way, the physical environment provides an immediate tangible focus for debate. The technique appears to offer the advantages of flexibility of research, group cohesion and a good environment for individual professionals to develop an understanding of other disciplines.

To some extent, the group trek approach can be seen simply as a pragmatic response to the conditions in Nepal where the majority of villages are only accessible by foot, the traditional society is used to catering for the foot traveller, a trek is more economical and safe if conducted in groups and there is nothing to do in the evenings other than chat about the day's work. The project's response has been to take these limitations and turn them to advantage. The office worker's trip to the field is no longer an under-valued burdensome task to be fitted in between more urgent matters; it has become a required and valued part of work which generates reports, programmes for action and other recognizable and respectable products. Another comparable technique is the *sondeo* (inquiry) developed in Guatemala. Five agricultural scientists and five social scientists work in pairs over a period of a week. Each day the partnerships are changed and each evening there is a group discussion (McCraken *et al.* 1988: 52). Similar techniques have been described elsewhere in which the multi-disciplinary team includes two or three of the local farmers (Lamug 1989: 74–5).

Other observers have described the 'mass walk' in which pretty much the whole population of a village walk on a tour of inspection with the researcher and so have a chance to discuss the issues *in situ*. Descriptions of mass walks smack of post-rationalizations of spontaneous events. Generally, a visit to a remote village will result in a mass walk, whether it is programmed or not. But where individual interviews are desired the spontaneous mass walk may make them impossible. This is not to say that mass walks or mobile meetings are not useful. It can be valuable to have the potential site of a new community project in front of one, along with representatives from all the households. If the visitor is being shown a completed project the walk often brings out the history of the project as different features along the way prompt anecdotes. This kind of process is particularly useful where there are linguistic problems. Mass walks are a category of event which one should welcome and exploit when they occur, but it would be arrogant of the researcher to programme for a whole village to give up a couple of hours to inspect a site. The mass walk cannot form part of a blueprint for a research programme but may nonetheless generate legitimate results.

A more overtly scientific approach to the walk is the 'transect' (Conway 1989: 78–80). Transects have long been used in botanical, soil and archaeological surveys in which a straight line is drawn on the map and samples are taken from spots along the line as translated to the field. The transect walk follows a similar approach although less emphasis is placed on the geometric purity of the straight line than on achieving a representative series of linked ecosystems. Transects are most obviously useful where large changes of conditions are experienced over a short distance, as occurs in mountain regions. Transect studies on mountain agriculture have been carried out by Conway in northern Pakistan (1989: 79). Walks were taken along the lines of the transects and interviews conducted along the way.

THE VILLAGER AS RESEARCHER

The principles of chance learning which apply to the intervener also apply to the intended beneficiary. He or she will not seek out information until he or she knows that it exists and is of interest to him or her. In the same way that the development worker needs to create contexts in which there is an increased probability of information emerging regarding the villagers' thoughts, so too must the field worker engineer situations in which the chances of a villager encountering a new idea are enhanced.

While children are obliged to go to school, learning cannot be imposed on adults. The learning of new ideas either has to be accidental or self-driven. There are three modes of learning:

- **Search**   The search occurs when the villager has identified a problem which is stated in a way which suggests a certain type of solution. He seeks out the relevant information because he has a particular reason for wanting it. 'The things we learn because, *for our own reasons*, we really need to know them, we don't forget' (Holt 1983: 134, original emphasis). The villager determines the subject and the circumstances for the search and selects the source which he considers competent to be able to provide an answer. The competent source may be the villager himself if he or she is going to carry out his search through experiment. If he or she has to seek the information from another person then there is an implicit recognition of the other's superior competence. This may or may not involve a loss of face.

- **Discovery**   Much of the information which we absorb we come across by chance. In many Third World countries it is common for people to hang about in the streets watching people work, maybe gathering in a group to watch a workman dig a hole, to make a repair to a pipe or lend a hand to help a neighbour pour a concrete slab. In some instances, this behaviour is part of a deliberate search for information, while in other cases it is accidental learning. At Centro Sinchaguasin, many of the people who went away having learnt something about building techniques had visited the Centre in order to buy tree seedlings. The discovery was by chance, but the situation existed where such chances could occur. By mixing a range of otherwise unrelated activities the opportunities for fortuitous accident were increased.

- **Instruction**   In the consideration of development education it appears to be generally assumed that the task is one of instruction.

Both booklet and course are predicated on the idea of a structured sequence of predetermined messages. Submitting oneself to instruction is potentially the least respectable of the three modes of learning. To be openly told that 'this is what you should do' and to accept it is to recognize one's subservient position, equivalent to that of a child before a parent or school teacher. 'Most of the time, most of us do not like at all to be confronted with someone who knows a great deal more about something than we do' (Holt 1983: 131).

In some circumstances to be seen to be receiving instruction can be status-enhancing, in the same way that going to a college of higher education can. Where a course has gained respect and maybe issues certificates to people who have completed it, instruction can be respectable. But even in these circumstances the actual process for adults of submitting to instruction in the classroom can be difficult and ineffective.

In the client–tradesman relationship, instruction takes on a different connotation. The client is instructing a tradesman on which tasks to carry out – not teaching him how to carry them out. In this case, instructions are the recognition of the client's respect for the tradesman's competence. In the last chapter the example was given of a builder who was employed to make an earth/cement/straw floor. The builder was being instructed to do something and within the respectable context of paid employment he was able to discover a new technique.

It has long been recognized in health programmes that waiting areas at clinics are potential learning situations. In practice, the reality often amounts to a few ineffectual posters – but opportunities do exist for demonstration kitchens, houses, and latrines, theatre groups promoting health messages and graphical material which can at least trigger curiosity. Wherever people gather and pass the time of day, idle chat is going on and information is being passed on and discussed. At Centro Sinchaguasin, the water supply which was installed for the tree nursery became the social hub of the neighbourhood where people came to collect water and to wash clothes. It had not been intentionally planned as a tool for dissemination and evaluation, but by its nature it became so. In another project such a natural social centre could be planned for and enhanced as a learning environment.

Search and discovery may work together. A successful search

leads to discovery while a discovery may initiate a search. If the intervener can create situations in which the opportunities for chance discovery are enhanced he may be able to trigger a virtuous circle of search and discovery. Situations may occur in which all three learning modes are employed. A visiting villager to Centro Sinchaguasin came across our demonstration composting latrine – he discovered it. Once his interest had been aroused he sought out more information as to how it worked, which he then communicated to his fellows. This resulted in the request for instruction in how to make the latrines in their village.

After the earthquake in Yemen, villagers discovered through observation that the damaged houses had failed most often at the corners. As reconstruction got under way people sought ways of strengthening their buildings and encountered the Dhamar Building Education project. Over a thousand builders enrolled for courses of instruction (Leslie 1987: 47). In these circumstances there was a place for instruction since it resulted from a prior legitimatizing process of discovery and search. The villagers were addressing their own problem.

THE PRINCIPLE OF MAXIMUM SERENDIPITY

Seek out situations which increase the chances of intervener and villager discovering something which the other knows. Do things together. Create a tangible focus and an understood context.

The need to establish opportunities for discoveries, for both villagers and interveners, is an extension of children's play; it almost doesn't matter what you do so long as you do something. If your first idea is a good idea then that's fortunate, but even if it is not it may provoke a response or suggest another potentially constructive avenue for play. The stone lines of Burkina Faso, described in Chapter 1, evolved from an intervener's first stab at a solution. The solution was abandoned, but in the process a dialogue developed in which both parties learned more of the other's knowledge and which finally provoked a sustainable and immensely valuable solution.

Part II
RECOGNIZABLE

Before anyone can evaluate and adopt an idea he or she has to understand what it is. Often an intervener and the intended beneficiary are talking at cross-purposes, with the intervener moving on to a consideration of details before the basic idea has been grasped. A new idea must fit into a person's existing structure of knowledge and have a name. Educational materials which are used to try and explain the idea must be comprehensible to the intended beneficiary.

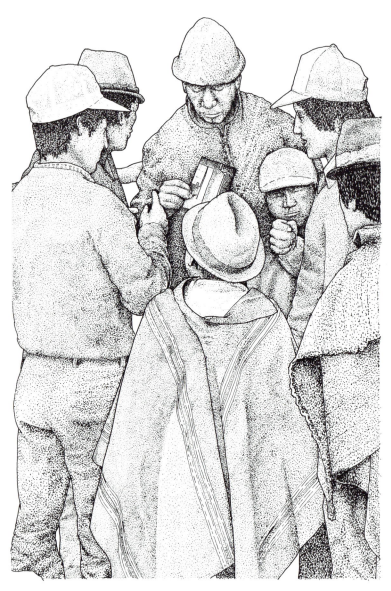

Figure 3   Discussing a promotional postcard
Picture postcards are an effective means of conveying simple messages. The postcard is a good discipline for interveners to make them reduce their ideas to a single recognizable image. Once the idea has been recognized it may be reasoned about and discussed.

4

TANGIBLE ENTITIES

In order to become acquainted with a thing we must consider
all of its prerequisites, that is, everything which suffices to
distinguish it from any other thing. This is what is called
definition, nature, essential property.

(Gottfried Wilhelm von Leibniz *c.* 1693)

In a study in Nepal, a communications worker, Ane Haaland,
interviewed mothers leaving a health education session at a clinic.
The object was to assess how many of the health messages had
been taken in. The results revealed that almost nothing had been
absorbed; the words and concepts which were being used in the
education session bore no relation to either the language or the ideas
which the mothers had about how the human body worked. They
could not even begin to consider the reasonableness of the idea
because they could not recognize a coherent idea which could be
subjected to reasoned consideration.

Other experiences from health care suggests that it is easier to
introduce expensive pills and injections in the Third World than
new practices of hygiene or nutrition. This may partly reflect the
promotional techniques of the pharmaceutical companies, but it
also has to do with the nature of the product – the tangible entity.
'Cure equals new pink pill' is a more readily assimilated and
attractive message than 'health is protected by washing your hands'.
Oral rehydration treatment has been one of the great success stories
of primary health care, yet it seems to be easier to promote the
sachets of ready-mixed powder than to persuade people to make
their own rehydration solution from salt, sugar and water. Similarly,
a tractor or a bag of fertilizer can appear to offer a comprehensible
and apparently instant solution to a problem. It is possible to desire,

buy or be given 'one of those'. The thing provides a focus for an aspiration and the context for the contemplation of a problem.

THE STRUCTURE OF KNOWLEDGE

In tackling any problem it is necessary to understand the component parts; what are the things of the situation. The natural division of parts will not necessarily be the same for everybody. The house-user's conceptual model of the house may be different from that of the builder; where the builder may see walls, a user sees rooms. Where an engineer sees a water tap, a local resident may see the social centre of the neighbourhood. If technical aid is about introducing ideas then it is useful to understand the structure of the existing body of knowledge into which one's proposed innovation will have to fit.

Consideration of indigenous structures of knowledge has been at the heart of anthropology for the last thirty years (for example Lévi-Strauss 1961). This work has tended to focus on social structures, particularly kinship, myth and status. More recently, researchers in development studies have been placing greater emphasis on uncovering indigenous technical knowledge and the ordering principles underlying that knowledge (for example Howes 1979: 5–11; Chambers 1983: 75–101; Chambers et al. 1989). Their experiments have demonstrated that villagers often classify their knowledge on different criteria to those of outsiders.

In one, now classic, experiment, a group of villagers were given groups of weeds, seeds or rice varieties in sets of three. For each set they were asked to say which two members were most similar and hence isolate that which was most different and then to explain the 'construct' or criterion underlying the choice. A set of binary constructs, such as high yield/low yield, was built up and used until a comprehensive evaluation of all the items had been carried out. In effect, it was a questionnaire which had been written by the respondents themselves. The villagers used constructs which related to practical characteristics associated with the plants, such as ease of clearing or medicinal uses. Some criteria were elicited which would probably have been missed in more conventional agricultural research. For example, one type of seed was found to be valued as a gambling chip. The experiments on weeds were repeated with groups of botany students and agricultural promoters. The botanists used a classic 'scientific' botanical taxonomy. The agricultural promoters, who were not trained in botany, proposed criteria which, though free of scientific jargon, were virtually identical to those of the botanists: such as root–non-root and round leaf–multiple leaf. This finding led to a debate among the promoters as to whether their way of ordering and presenting information was acting as a barrier between themselves and their intended client population (Richards 1980: 187–91).

In a highly controlled industrial production process, the techno-logical knowledge of the individual worker has no influence on the nature of the outcome. It is not necessary for the worker to understand either the purpose or the principle behind the product. The worker on the production line is simply required to act as a machine, performing a task reliably and without question. The fibre-cement tiles, discussed in Chapter 1, can be successfully made by relatively unskilled workers if they are prepared to blindly follow instructions from above. In the modern building process, an engineer or an architect conveys information to a builder largely through drawings. The process of design is divorced from that of realization. The drawing, not the building, is the result of the decisions of the design process. Once the drawing is done, provided the builder executes his task correctly, the nature of the finished building is a foregone conclusion. The drawing is the focus of attention; the building is the by-product.

In societies in which design drawings are not used, the building process has to proceed in a series of self-sufficient steps, each of which must make sense in terms of the knowledge of the builder. Each step must be capable of communication through words and gestures:

- 'I want to build a room seven metres by five, pointing this way.'
- 'Put a door here.'
- 'Build me a brick wall between these two points.'

Each of these building tasks is related to a physical entity and each is amenable to independent change. Some of the entities which make up a house are physical building components, such as bricks, while others are complex assemblies such as kitchens, fireplaces and entrances. In some societies, the instruction from client to builder of 'build me a house here' may be sufficient. The word 'house' may convey a precise model whose form, size, and materials are under-stood by everyone in that society (see Rapoport 1969: 4–5; Alexander

1964: 48). In northern Pakistan, until recently, it was common for the village carpenter to charge a fixed rate to build a roof since everybody knew what, and how big, a roof was.

In strongly traditional societies it is possible for the researcher to define indigenous architecture in terms of the recurring configurations of space and their functional associations. The significance of culturally defined structure may be both obscure to the outsider and highly localized. Two examples from the traditional domestic architecture of the Karakorum mountains of northern Pakistan illustrate how finely tuned the structure of meaning in buildings may be.

In the western Karakorums, the conventional house type is often referred to by the Urdu word for stove – *baipash*. The *baipash* is essentially a single room, often with a store attached, and sometimes an entrance lobby. The principle concern is how to keep warm during the long, cold winters. All domestic activity, including cooking and sleeping, takes place in the single space around the fire. The tradition of intense living in a confined space has generated an architecture reminiscent of the inside of the cabin of a small boat; every corner has its own distinct purpose, name and form.

The principle mechanism used for demarcating space in the *baipash* is to break the floor up into a number of platforms. The platforms are each assigned names and functions; each of the platforms defines a system of space, activity, and has a name. For example, the platform nearest the door, called the *loop-raj*, is the male sleeping platform. On special occasions, such as weddings and funerals, the upper end of the *loop-raj* is where the political leader sits, while the lower corner is for the religious leader. At other times the upper corner is reserved for sick people, while the lower corner is for those who are dying.

The second example is from Baltistan in the eastern Karakorums. In discussions with householders nine different words emerged which in English would mean kitchen:

aserkhung	the modern 'white' kitchen;
dukang, zankhung	a spring kitchen;
abierdouk, zverduk	the summer kitchen;
tseley	a summer kitchen as a corner to a veranda;
balti, zankura	a kitchen for spring, autumn, and mild winters;
katza (or katzi)	the kitchen for deep winter.

Not all of the names apply in all parts of Baltistan, and in some instances one name can have different meanings depending on the location. For instance, the *katza*, is a windowless room on the ground floor of a two-storey house or, occasionally, is found in a basement. In much of Baltistan, *katza* means fodder store. Since fodder not only makes good insulation but also generates heat as it decays, the *katza* is generally the warmest unheated room of the house and makes a sensible refuge at times of extreme cold.

In some places there will only be a few days of the year when the climate necessitates retiring to the *katza*. In a mild winter there may be no need to use it at all. Here the family will simply be camping in the *katza*, and so the *katza* will be perceived as primarily a fodder store. In higher altitude communities it may be necessary to live for two or three months of the year in the *katza*. In such a case the fodder is accommodated in an adjoining room, and the room will be fitted out with shelves, cupboards, decorated timber-work and a hearth. It may also be divided by a low wooden wall in order for animals to occupy half of the room and so help to heat the room with their body warmth. In these circumstances, the word *katza* unambiguously means 'the winter room'. At the other extreme, in lower Baltistan, where the climate is less harsh, there appears to be no tradition of using the *katza* to live in and the idea is greeted by the inhabitants with apparent incredulity.

The first example, the *baipash*, illustrates that even in a one-room house there can be a high density of meaning and that the functions of the various parts may alter according to the occasion. The second case shows how the meaning of a single important word can vary across a small geographical area among people of the same basic culture and language. Both examples appear to suggest that any outsider attempting to introduce a new technology into such a situation is likely to meet problems. Even exhaustive anthropological studies would be unlikely to reveal the full depth and cultural significance of the indigenous technological knowledge.

The above examples are both from isolated communities in which the indigenous way of life can be considered to be still relatively 'pure' in the sense that the influence of modern urban culture has been limited. However, such examples are becoming quite scarce. In Ecuador, for five hundred years an imported language and culture was imposed onto a traditional society. The process of modernization is more advanced than in northern Pakistan and consequently the indigenous structure is more diffuse. In Ecuador, and many other countries, the kind of tidy classification which distinguishes each type of platform in the *baipash* is not possible in the modern home.

COMMON STRUCTURE

To enable the intervener to act, it is useful to identify common fundamental structuring principles which apply both in situations with an apparently weak indigenous structure, such as in Ecuadorian housing, and in those with a bewildering density of structure. The work of Lévi-Strauss and his followers primarily sets out to demonstrate that despite the rich variety of superficial social phenomena there are 'deep structures' of ordering principles which cut across cultural boundaries.

A study of the classification of colours in a variety of widely dispersed cultures revealed more or less consistent linguistic boundaries between the principal colours and a common pattern to the hierarchies of colour complexity (Berlin and Kay 1969: 3). Languages with increasingly greater complexity were shown to progress through an approximately predictable sequence of colour terms. In languages which only distinguish between three colours, the terms were invariably found to correspond to white, red and black. If there was a fourth term it could be either green or yellow, whereas if there were six colours they were consistently found to correspond to white, black, red, green, yellow and blue.

In architecture, there are entities which, despite the wide scope for detailed variation, are consistent across cultures. The door, column, pitched roof, beam and brick are examples of widely understood categories of built object. Such elemental entities have meaning even in the cultural chaos of rapidly modernizing states. Doors and columns are often endowed with a wealth of cultural significance, yet however complex the layers of added meaning they are still recognizable to people across the world as *a means of entry* and as *a means of structural support*. There are also universal features in the task of building which are derived from the intrinsic characteristics of a building material. Many of the tasks faced by a stone mason in Bolivia would be recognized by a mason from Tibet. Similarly, the process of conceiving of built form has universal characteristics which emerge from the intrinsic nature of space. For example, in spite of the rich variety demonstrated in the history of architecture there are still only two common ways of creating a room: either a cell is added or an existing one is subdivided.

In the modern building process the steps of addition and sub-division take place on paper and in the architect's head. In many architect-designed buildings the underlying additions and sub-divisions may be lost under embellishments and rationalizations. The process of rationalization – in which one element is made to serve several purposes – is, arguably, the essence of the design skill. To the user and the builder of a modern building the derivation of a space and its enclosing elements may not be clear. The brick-layer is concerned with placing his bricks where the drawing has indicated and following laid-down instructions whose purpose is obscure. The wall may need to include steel reinforcement or holes for drain-pipes which bear no obvious relation to the wall's fundamental purpose of enclosure and support. It is an architecture of prescription which works in the context of control. Where drawings are not used the householder, together with his builder, are designing the building on the ground in the context of earlier building acts. The elemental process of spatial addition and subdivision is directly related to that of building. A common characteristic of much vernacular architecture is the intelligibility of its additions and subdivisions.

The technical aid intervener must suggest technologies which make sense to the user–maker within the context of the step-by-step sequence of understandable acts. For instance, in the 1987 earthquake in Ecuador, many of the rammed earth buildings were damaged because of the poor connections between walls. The traditional rammed earth mould is a straight box and so corners are made rather like brickwork with one layer lapping over the layer below. A typical brick is twice as long as it is wide and so the overlap covers half the area of the brick. A rammed earth mould tends to be relatively long and thin, typically the length is about four times the width. Consequently, relative to the size of the mould there is a small overlap and there is a disproportionate accumulation of joints at the corners. Since in an earthquake stresses effectively accumulate at the corners of buildings, the points of greatest weakness are then coinciding with the points of greatest stress. In order to combat this problem, an NGO suggested the introduction of two new technologies: an 'L' mould and a 'T' mould. The 'L' mould helped to build stronger corners while the 'T' mould strengthened the junction in which an internal wall butts up against an external wall.

In practice, the reconstruction programme concentrated on the

'L' mould since all houses have four corners whereas only larger houses have internal divisions. However, it also seemed probable that the 'L' mould was perceived by the builders as having a coherent place in the construction process while the 'T' mould did not. The normal way to build a house in that region is to build a rectangular box and then to subdivide it, maybe at a later date. The processes of building the overall box and demarcating the rooms are seen as two distinct tasks. The 'L' mould does not disrupt this pattern, it allows an existing task to be done better. The 'T' mould requires the builder to raise the internal walls at the same time as the external walls. It demands not simply a change of technology but a change in the mental construct of the building. No longer is the house one entity divided up into subservient ones, but rather it has become a single multi-celled entity.

In Yemen, the staff of the Dhamar post-earthquake building reconstruction programme also suggested to builders that they raised all of the masonry walls together to ensure that the internal walls were bonded to the external walls. This technique provided a significant improvement in the strength of the house at zero-cost, yet it suffered the same conceptual problem as the 'T' mould in Ecuador. In the event, it proved harder to promote this practice – which involved a re-conceptualization of the house – than to promote a tangible entity such as a concrete ring beam. The ring beam is like an accessory, it adds on to the existing structure of entities. Similarly, the 'L' mould is like an upgraded spare part; it permits you to do an existing job better without disrupting the mental construct of the larger entity – the house.

In northern Pakistan it is now common to find that people have put a damp-proof course around the top of their stone walls. In English a damp-proof course is generally known by its initials, d.p.c. The non-English speaking builders of northern Pakistan use the same initials to describe the entity of the damp-proof course without knowing what the full name is. The initials have turned into a new word, *deepeecee*, which describes that recognizable object. The important thing is that it has a name, any name, however nonsensical; without a name no one can be told to make one.

REQUISITE MAGIC

To be told that one's crops could be improved by taking more care of the dung from the animals or to hear that one's house could be made stronger by taking more care over how the stones in the wall are laid is to be told that what one was doing before was wrong. Nobody likes to learn and to accept that they were wrong. To be told by an outsider, however nicely, that one is incompetent is doubly shameful. A new thing, 'ingredient X', provides a respectable excuse for change. Everybody likes a gadget which seems to do something wonderful and almost magical. In the developed world people are fascinated by the latest examples of technical wizardry and the same is often true in the villages of the Third World.

The 'green revolution' was characterized by precisely this kind of miracle cure. Its appeal lay in clearly-understood benefits deriving from the use of a new variety of seed. Its problems arose from an under-perception of the other inputs required in order to work the miracle. The sachet of Oral Rehydration Salts and the painful injection inflicted by someone in a white coat both share the quality of mystery. In Ladakh, the most striking spontaneous technological introduction of recent years is the pressure cooker. It is a technology which is eminently suited to high altitude communities. In Leh, the capital of Ladakh, water boils at just 84°C, hence it takes more time and more fuel to cook food in water than it does at sea level. With a pressure cooker, cooking time is drastically reduced. The benefit is clear and the source of the benefit is tangible, readily transported, and has a name. Development programmes in Ladakh designed to introduce improved cookstoves have, like so many others elsewhere, failed to capture the imagination of the intended users. Apart from any practical problems, the benefits are marginal and the improved technology is not in the form a clear 'new thing', but rather in the subtleties of the precise shape of combustion chambers and flues.

The new ingredient or gadget needs to be significantly different from the old one and to result in a distinct improvement if it is to be perceived as a real change. When people make a change in their way of life they want the result to be something other, not just a mildly enhanced performance. The Trombe walls in Ladakh, mentioned in Chapter 1, are certainly distinctive in appearance, but the result is not striking. In contrast, the locally-developed glass room can create luxurious heat on a winter's day (see Figure 2). The concrete latrine slabs in Mozambique, also mentioned in Chapter 1, were designed as thin shallow cones which, because of their shape, did not need steel reinforcement. The result was a lightweight, low-cost and visually-distinctive technology. An alternative approach might have been a 'How To Make a Latrine' manual or course. But such a route would have lacked the missing magic ingredient, the reason why it couldn't have been done before.

A contra-seismic technique which has been attempted in a number of projects, without much success, is the concept of roof stiffening. A roof structure has the potential to tie the walls of a building together and to provide resistance to the horizontal forces induced by an earthquake. The project in Yemen and a number of projects in Ecuador advocated means by which conventional roof structures intended to take the weight of the roof could be adapted to resist the horizontal forces as well for little extra expense. In Yemen, where roofs are flat, plywood is often used to create a deck on top of the rafters on which a waterproofing layer may be placed underneath the traditional earth covering. By simply ensuring that the plywood is well nailed to the rafters, using more nails than otherwise, the roof structure becomes a horizontal structural element which, if adequately connected to the walls, will considerably strengthen the whole building. The technique is cheap, simple and effective, and unlike the 'T' mould does not disrupt the existing set of entities. However, it is weak on two counts:

- It does not involve a tangible new entity. Unlike a ring beam, it is not possible to 'build one of those'. Equally, unlike an 'L' mould, it is not possible to build *with* 'one of those'.
- It demands a reconceptualization of the purpose of the existing entity – the roof structure.

There are many examples of potential improvements which are not in themselves entities. Some are simply desirable options, while others are essential. In health care, practices of improved hygiene cannot be replaced by any cunning technological device. Similarly, a balanced programme of housing improvement cannot be achieved solely through entities. If earth-construction techniques are to be used, practices concerning maintenance and repair are essential. While recognizing that the tangible entity lends itself to promotion and adoption, means have to be found by which improved practices may be introduced.

In Chapter 1 it was suggested that the most useful technologies would be those Big Ideas which had no background deficit and no essential practice while having ample opportunity for good practice. Yesterday's good practice is today's Big Idea. Once reinforced concrete has become an established part of the context, new technologies may be learnt for the better making of concrete. Knowledge regarding how to fix roof sheets, optimum rafter spacing, how to build a plumb wall, how to repair a damaged water pipe, how to lock a pivot door and how to improve the quality of building soil are all in the realm of potential indigenous innovation.

In some cases the ancillary practices may not be directly concerned with the execution of the introduced entity. For instance, a hinged door in an earth wall results in damage to the wall while a pivot door puts less stress on the surrounding wall. This suggests a corollary: if we want to introduce a new practice its chances of adoption might be enhanced if it were introduced in the wake of a convincing new entity. The introduction of a pivot door may act as an enabling technology to an initiative for improved finishes.

The 'water level' introduced in the Yatenga project in Burkino Faso is an example of such a technology. It is a tangible and ingenious device which, through its use, provides the opportunity for a host of good practices. Methods of cultivation which were already known about could be discussed and encouraged in the name of getting the best out of your new gadget. If a concrete ring beam is a new idea and one is told that a well-made ring beam requires a well-bonded wall beneath it in order to be effective, then there is both a motivation and an alibi for a change of practice.

In agriculture, development programmes commonly advocate practices such as composting. Again, these programmes often encounter problems since there is no visual focus for the intended beneficiaries. When such programmes switch from a general consideration of the process to a discussion of physical objects, such as compost bins or composting latrines, then there is not only a greater chance of take-up but also of fruitful dialogue around the use of an actual thing. In some cases the physical manifestation of a process may not be an entirely clear entity but can still be expressed in concrete form. For instance, the principle of inter-cropping when presented as alley-cropping takes on a physical configuration which can be seen, conceptualized and discussed.

The 'L' moulds were introduced for reasons of seismic strengthening, but they provided a catalyst for elevating rammed earth to being a visibly higher quality product. In the Hunza valley, the 'L' mould was also introduced. As in Ecuador, rammed earth is a traditional but rustic technique. The earthquake risk in Hunza is lower than in Ecuador so, more than anything else, the 'L' mould served to introduce the idea that rammed earth could be used to produce buildings of quality. Straight moulds were used to produce walls of a significantly higher quality than before, and cleanable surface finishes for earth were introduced. A parallel may again be drawn with health care. A vaccination programme, which involves a clear, painful – and thus powerful – entity, is a way of reaching mothers and children with monitoring programmes and other less immediately attractive messages. The tangible entity is the hook.

THE PRINCIPLE OF TANGIBLE ENTITIES

A 'thing' is easier to grasp than an idea. Where possible, identify a technology which may be presented as a tangible entity. A new 'thing' provides a reason for change. Improved practices may be more readily introduced in the wake of a tangible entity

The implication of the Principle of Tangible Entities is *not* that one should not bother to promote practices but that:

- If one has a number of options then, if all else is equal, a solution which involves a tangible entity is preferable to one which does not.
- If one needs to promote a practice then one should look for opportunities for promoting it in the wake of a convincing new entity.
- The entity should produce the minimum disruption to the logic of the existing assembly of entities as perceived by the user.

5

CLEAR VISUAL MESSAGES

To an old woman in Luapula, Zambia, I gave a picture handout of a woman breast-feeding her baby. The handout was A4 in size and printed on glossy art paper. We asked the old woman what she saw on the paper. She seemed not to understand the question. Instead, she lifted the picture to her nose, smelling it and feeling its smooth surface with her fingers. It was the intense whiteness of the paper, its straight edges and sharp corners which attracted her.

(Andreas Fuglesang 1982: 145)

Visual material is commonly used to help convey new technical ideas. While literacy is regarded as a key social indicator of development there are no commonly-used measures by which other abilities to interpret visual information may be judged. The written word is just one of the ways in which we use our sense of sight to absorb messages. Interpreting pictures and three-dimensional objects requires skills which, like reading and writing, have to be learnt. When we look at a scene or a printed page, we do not necessarily all see the same thing. A diagram of an electric circuit may be informative to an electrical engineer while to the rest of us it is meaningless. In other cases, the nuances of understanding and meaning are more subtle and diverse; an architect, a builder, a cleaner and a burglar are liable to see the same building in quite distinct ways. For the large numbers of rural people who are illiterate, or of very limited literacy, the visual information which they absorb necessarily comes in a non-verbal form. Even for those who are literate, much of their visually acquired information will also be non-verbal.

THE PROBLEMS WITH BOOKLETS

In Latin America, due to the work of Paulo Freire, consideration of 'popular education' techniques is perhaps more advanced than elsewhere, and numerous methods including drama and participatory workshops have evolved (Freire 1972). In health education particularly, many of these techniques are increasingly widely used. However, in much technical education the predominant methods of dissemination remain the traditional ones of the training course and the booklet. While it is hard to generalize about training courses, certain recurring problems may be identified concerning booklets.

A typical development education booklet consists primarily of drawings, but it will also have captions. The use of captions poses a double problem: generally, the drawings cannot be understood without the captions and the captions cannot be understood without the drawings. The illiterate villager is disadvantaged both ways. He cannot use the booklet without the assistance of a literate friend and if a friend is found to read the text aloud the text does not, in itself, make sense; the captions strung together are not coherent prose. In one booklet dealing with house construction in Ecuador, the house was anthropomorphized. The result was text which rather than reading 'this is how to put a roof on a house', became 'this is how I can put a roof on myself'. This is liable to confuse more than it clarifies.

In a one-to-one situation the villager may be able to get sufficiently close to both hear the text and see the drawings. Even then, since there may be several captions to a page, the listener may not be clear about which caption is being read for which picture. The more likely situation of one literate, or semi-literate, villager reading to a group of his fellows is precluded since they cannot all see the pictures in order to make sense of the words. Since a booklet may be twenty to forty pages long, if not more, it may be hard to persuade someone to read through it all, and virtually impossible to find someone to repeat the performance.

Since the intervener's task is to convey information, it is often felt that the more information which is presented, the better the job is being done. In practice, the reverse may be true. A booklet may defeat its own object with an excess of information. When a commercial manufacturer promotes a new product he will start by using a single attractive image which portrays the product as desirable; he does not sell the product through the medium of the instruction manual. In Chapter 1, the distinction was drawn between the deficit of background knowledge, the Big Idea, the essential practice and the good practice. In a booklet, all four types of information tend to be mixed up. If one strictly followed all of the advice in a technical education booklet, a good result might follow but the critical Big Ideas are lost among the secondary details. A surfeit of information is boring, threatening, incomprehensible or all three.

The comprehensive booklet or manual can create the impression that it contains and defines the technological package of correct solutions. Most building education booklets are of the 'How To Build a House' type rather than in the 'More Ways To Make a Floor' category. The 'How To' booklet implies both that there is a single definitively correct way to build a house and that the villager is in need of instruction. The 'More Ways' type simply offers some additional alternatives to the builder's repertoire, preferably accompanied by all the respective pros and cons. The 'How To' booklet links a number of disparate technologies together in such a way that the user may feel he has to accept or reject the package as a whole. The booklet will show designs for a foundation, a wall, a roof and every other part of the building. What is not made clear is whether the wall may be used on a different foundation or under a different roof. Since, in such a booklet, every part of the house is mentioned, pressure is put on the intervener to come up with something innovative to say about every aspect of the construction. Inevitably some of these 'innovations' are liable to be less successful than others and may discredit the whole package. A booklet can be in circulation for a number of years and, if the booklet is indeed doing its job of communicating ideas, it may be propagating errors. Fortunately, there seems to be very little evidence of booklets actually having their intended effect.

Publications are usually the only tangible and lasting evidence of development education activity. Since development aid is a competitive enterprise, there is pressure on donors, development organizations and individual development professionals to use publications as proof both of their productivity and innovatory capacity. A booklet which is ostensibly aimed at semi-literate villagers is likely also to find its way on to desks in Washington, university libraries in Europe and the offices of one's rivals. The human desire to demonstrate how much one knows again creates a tendency to maximize dubious information rather than reduce information to a useful and reliable minimum.

Occasionally, some or all of these pitfalls are avoided. There are booklets in which the content has been well thought out, the target group has been kept firmly in mind and the use of text in conjunction with graphics more carefully resolved. Increasingly, in preference to booklets, projects are using flip charts in which large-scale graphics are shown to a group by a trained field worker who reads a text which has been designed from the outset to be read aloud. Yet, even in the better designed publications, the problem of interpreting pictorial material remains.

READING GRAPHICS

When drawings are presented to villagers and their opinions sought as to what the drawings portray, it often emerges that the designers of booklets and posters assume an understanding of graphic conventions which does not actually exist. Several common barriers to understanding appear:

- The use of unfamiliar codes
- Cartoons and anthropomorphism
- The portrayal of abstract concepts
- The passage of time and sequence
- Abstractions of physical phenomena
- Maps, plans and sections
- Extraneous detail
- Designer graphics.

Codes

Codes such as the tick and the cross for right and wrong, which to urban intellectuals are so deeply ingrained as to be hardly perceived as codes, may be unknown to the target audience. In a study in Pakistan, villagers shown drawings including a tick and a cross interpreted them as such things as a ceiling fan, an aeroplane, an Urdu seven and a bent piece of metal. Even the few people who were familiar with the good/bad connotations of the tick and the cross, generally did not relate them to the corresponding drawings illustrating good and bad practice (Dudley and Haaland 1993: 43). A study by UNICEF in Nepal reports that the drawing reproduced in Figure 4 was shown to villagers who were asked what they thought the drawing was trying to teach. Of the 410 respondents only 3 per cent understood that the drawing meant 'breast-feeding is better than bottle-feeding' (Haaland and Fussell 1976: 35). Only one respondent appeared to be making the connection between the tick/cross code and the drawings presented below. Conversely, in Yemen, among builders who had worked on urban building sites, the tick and the cross was found to be widely understood and could be successfully employed in building education materials (Leslie 1987: 47).

Figure 4 The tick and the cross – an unsuccessful use of a tick and a cross to convey a health message in Nepal

Cartoons and anthropomorphism

Use of cartoon characters is often taken by the development community as indicative of a professional piece of work. But the ability to understand cartoon characters stems largely from an urban experience of television and newspapers. When cartoon characterizations of good and bad houses were tested with villagers in Pakistan the result was bewilderment. Some people would say that the drawing (Figure 5) showed a house 'but there is something wrong with it'. Others would interpret the eyes as electricity meters and the mouth as maybe a boat or a window. One man suggested that there was a house with someone in it. But the idea that the

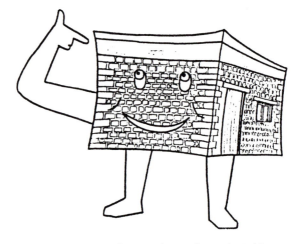

Figure 5   Cartoon house – happy house in Pakistan

object could be simultaneously a house and something alive was incomprehensible. Other cartoon conventions such as speech and thought bubbles can be equally confusing. In Nepalese villages the speech bubble on a drawing was sometimes perceived as being a large clove of garlic (Haaland 1984: 8). Cartoon characterizations which are made in order to be fun, informal, expressive and friendly, can be confusing and are based on the mistaken assumption that illiteracy is equivalent to a childlike mentality.

Abstractions of concepts

Educational materials often attempt to make use of analogy and metaphor to convey a concept. In the Pakistan study several attempts were made to convey the structural concept of a ring beam. A typical example was that shown in Figure 6 which portrays the beam as a belt tying the house together. None of the respondents understood this message although many recognized both the house and the belt. A common response was 'we don't build like that around here'. As with cartoons, the idea that a graphic image could be conveying anything other than the literal truth was alien. In Ecuador, a series of booklets about building technology aimed at peasant farmers was produced by the Government in collaboration with a United Nations agency. In one of the booklets the illustration

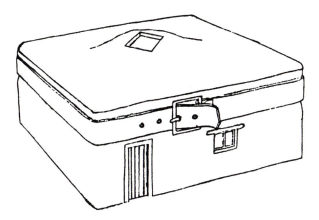

Figure 6 The use of analogy – using a belt as an analogy for a
strengthening ring beam

Figure 7 Abstract concepts – abstraction of a philosophical concept
relating modern and indigenous values

reproduced at full size in Figure 7 was used to represent the
choice between indigenous and imported technologies. The scale
and the level of detail alone made the drawing hard to read but,
even with improved graphics, the level of philosophical abstraction
required to interpret the drawing correctly is beyond that which
could reasonably be expected.

Sequence

Despite the problems with expressing abstract concepts, Gordon
Conway and others have demonstrated how diagrams may be used

by villagers to express the concept of time in relation to things like the agricultural cycle (Conway 1989: 77–86). However, his work was done in the context of small workshops in which the abstractions were built up by the participants themselves. In the Pakistan study several attempts were made to convey simple time sequences, none of which were understood. When three pictures of a single building at different stages were shown they were seen as three separate and unrelated buildings. The idea that an image on a piece of paper was conveying the passage of time could not be grasped. In educational materials sequences to show progressive operations and pairs of before-and-after images are common conventions. The 'How-To' manual is usually built on the generally-untested assumption that people can relate one drawing to another in a sequential way.

Abstractions of physical phenomena

Drawings are sometimes required to portray invisible physical phenomena such as sound and movement. Lines or arrows will indicate sunshine or wind; horizontal lines and clouds of dust suggest that a figure is running. Again, such devices are a code which has to be learnt. The apparently straightforward drawing of a man cutting wood in Figure 8 was shown to villagers in Nepal during a UNICEF study (McBean 1989). Only 38 per cent of the respondents understood what the drawing showed.

Figure 8 Movement – Nepalese drawing for a wood-cutter

Maps, plans and sections

Map-reading skills are often absent in relatively highly educated people. For many villagers maps are completely novel, though, as with time sequences, participatory workshops can make good use of maps when the idea is built up slowly and collectively. Architectural conventions such as plans and sections are also forms of abstraction. Plans are becoming more widely understood as people are exposed to urban building practices. Even for those who have not been involved with urban building, a plan does have some physical reality since a building may be marked out on the ground. Similarly, an elevation of a facade may well be understood since it is a 'picture' of the building. Sections are pure abstraction. One never sees the section of a building. Even among quite experienced builders, a section on an architect's drawing will often be ignored since it is not understood. The example from the Ecuadorian booklets of a section of a building in Figure 9 uses a number of sophisticated conventions such as 'hatching' to represent earth and arrows both for indicating dimensions and as a link between part of the drawing and a label.

Extraneous detail

In the Pakistan study, various ways of depicting the people in the drawings were tried. Attempts were made to portray the people in local dress which in that area, Gilgit district, includes a distinctive

Figure 9 The use of cross-sections – a cross-section and other codes in a building education booklet

woollen hat. We asked whether the drawing portrayed a good or a bad builder. A number of respondents declared that the man was from the neighbouring district of Chitral and was thus clearly a bad builder. To an outsider, the Chitrali hat is virtually identical to that of Gilgit, yet to the locals something about the way in which it had been drawn conveyed the impression that it was Chitrali dress. The hat, intended to create a sense of local relevance, was actually alienating.

Similar results, from Nepal and elsewhere, have highlighted that extraneous detail can be a major source of confusion. If someone is having trouble understanding a drawing they will search for recognizable parts or try and find meaning where there is none to be found. Conversely, we found that reinforcing relevant detail helped greatly in the understanding of drawings. With drawings intended to convey the well-built house, people would pick up on very fine detail such as the way in which the stones in the wall were drawn. It is not that drawings should be simple but rather that they should make literal sense. When walls were left blank they were interpreted literally as smooth concrete walls and hence as strong walls.

Designer graphics

Harder to pin down are stylistic games which designers of all kinds have learnt to perceive as the elements which elevate the mundane to a piece of design. In the Ecuadorian booklets, all the pages have a standard border which is a typical stylistic device designed to tie a booklet together into a coherent whole. Figure 9 shows how the designer has allowed the graphic images to cross the border in an 'interesting' and 'informal' way. I would suggest that such apparently innocent games add nothing to the quality of the communication, while running the risk of confusing the inexperienced reader of drawings.

The ability to read graphical conventions is a skill which, like reading and writing, may be learnt. The 1989 UNICEF study in Nepal showed how – over a number of sessions with villagers, exposing them to images and explaining their meaning – the interpretative skills of the villagers could be improved (McBean 1989). The study also revealed differences in the ease with which different graphic conventions could be learnt. The ability to interpret representations of objects and perspective drawing increased considerably. Comic strips showing a chronologically linked series of images remained poorly understood, as did representations of motion. Even after six workshop sessions, less than half of the 240 respondents could understand the notion of representing movement in drawings such as that reproduced in Figure 8. In contrast, when the villagers were shown colour photographs of simple objects at the first session, 93 per cent of the people could identify the objects.

Serious study of indigenous skills in visual interpretation and their relevance for development communication is still in its infancy. However, the problems outlined above suggest some guidelines for making educational materials effective:

- **A clear target** The presentation should be aimed at a specific target group, such as village householders or local field workers. If it can be used by other people then that is a bonus, but nothing should be done which reduces its efficacy in communicating with the target group.
- **An independent technology** The presentation should not create any illusory linkages between sets of technologies. Each technology should stand or fall on its own merits and be seen simply as an additional option.
- **Essential information** The amount of information conveyed should be the minimum for conveying the message. Additional information should be *reinforcing* rather than extraneous detail. If there is any text it should be written in simple and coherent prose which can be read aloud independently of any graphics.
- **Literal representation** The use of visual codes should be kept to a minimum. Graphic representations should be as literal as possible, and cartoon caricatures generally avoided unless there is solid evidence that the target population is used to cartoons.

THE USE OF POSTCARDS

Benedict Tisa, a communications consultant, observed that when he received a postcard in his mail in the United States, the postman would sometimes read the card and pass comment. Later, when working in Bangladesh, realizing that a delivery of mail is more of an event in a Bangladeshi village than a New Jersey suburb, he decided to experiment with social curiosity as an educational tool. He produced cards with line drawings of a design for a water filter with a few written captions. He then posted the cards to field workers in other villages having previously asked them to watch for any reaction. To get to the field workers' houses the cards had to pass through several hands and in the process people saw the cards and their curiosity was aroused. Several people came to the field workers and asked them to explain the illustration and the meaning of the words. In this way the villagers were encouraged to approach the field worker rather than the other way around (Tisa 1984: 25).

In rural Ecuador there is no such system of village mail but we had noted that when we had given some colour snap-shots of technologies to a local village leader he was still using them several months later to explain the technologies to his neighbours. We decided to produce postcard-like information cards which, like Tisa's cards, would have the advantage of brevity and one single message. We produced a set of eight cards. On one side of each there was a colour photograph of a building technology while on the other was a short piece of explanatory text. The text consisted of a title, a statement of the problem and a description of the solution. When the cards were distributed it was often the case that a small group would cluster around one literate man, typically the village president, while he read the text aloud. The text was designed to make sense on its own. Once the reading was complete the card would be turned over so that all could see the photograph, and a discussion would ensue. Sometimes this discussion would slip from Spanish into the indigenous *Quechua* language, which may have helped those less confident of their Spanish to understand the idea. The fact that conversations took place in *Quechua* suggests that the interest was in the technology rather than in trying to please the foreign development worker.

Despite the apparent success of the postcards in stimulating discussion, several problems emerged. One of the cards dealt with a way of improving limewash by adding carpentry glue to the mix. The glue both made the painted surface stronger and eliminated the lime dust which tends to come off a limewashed wall. The photograph showed a man applying the limewash to a floor. The ingredients, notably the carpentry glue, were on the floor next to the man. The man in the photograph was someone who helped in the tree nursery and since, on that day, the sun was strong he had put on a hat. The hat which he had picked up was of a type normally worn by indigenous women. This had not bothered him since all he wanted was protection from the sun. The consequence was an image in which the most striking feature, for many of the indigenous villagers, was a woman's hat on a man. When the card was distributed it was frequently observed that the conversation focused on why the man might be doing such an extraordinary thing and people would come to us for an explanation of the meaning of this.

Whereas in a drawing an artist might add some feature for the wrong reason, in a photograph the error may be that of failing to notice and change some apparently inconsequential detail in the scene. In the Nepal studies, and elsewhere, it was often found that comprehension of a black and white photograph could be significantly increased by cutting out and discarding the background (Haaland and Fussell 1976: 12–17). When dealing with building technologies, we felt that the background context would often be important; a man painting a floor is only comprehensible in the context of a room. By using colour photographs, the possible confusions could be minimized. In our case, the problem was not one of an incorrect reading of an image but one of confusion caused by the correct reading of objects which were then seen to be in apparently incongruous conjunctions. When we produced a second set of postcards, we took more care in considering the significance of every aspect of the finished photograph. Any superfluous element in the background was removed before the photograph was taken.

The text on the back of the cards was done in hand-written capital letters. This is a technique common to many booklets and other educational materials. Handwriting is considered friendly and informal and a draughtsman can produce neat capital letters more easily than a combination of upper and lower case letters. Although I have no hard evidence, I now believe that this approach was mistaken. I do not believe anybody is lulled into a sense of friendly informality by printed handwriting as opposed to printed typeset text. Furthermore, the combined use of upper and lower case letters has evolved over two millenia because they have been found to be easier to read. Books for children and other poor readers generally use large, well-rounded, well-spaced typeset letters which generally have serifs (characters with serifs have short cross-lines at the end of the main strokes of the letters, unlike this text, which is sans-serif). It would seem sensible when producing material for semi-literate villagers to build upon the large experience built up in literacy training. In our second edition of cards we adopted typeset text.

The format of a photograph on one side and text on the other produced the inevitable problem of having to flip from one side of the card to the other. In the second edition of cards, a space was reserved at the bottom of the photograph for a title and a description consisting of one short sentence. Further information was still to be found on the reverse of the cards, but the essence of the idea was on the front. This format imposed a valuable discipline on us, the designers. The Big Idea had to be conveyed in a single image, a title and a single sentence. The title was particularly important since, as was mentioned in the previous chapter, people have to be able to give a name to something before they can aspire to own one. We realized that the titles on the early postcards were not all achieving this necessary clarity. The postcard which showed doors which could be hung on pivots rather than hinges was entitled 'the door without hinges'. In the second edition this became, more positively, the 'pivot door'. Where a simple name could not be found names were made up. In the case of the improved limewash, the first card had been entitled 'the finishes'; in the second edition it was *pintatierra*. The word is a contraction of the words for paint and earth and is presented like a trade name. Similar things have been done elsewhere; earth stabilized with cement is often referred to as *terrocemento*.

The postcard is not a panacea. Like booklets, they may be produced with an eye on one's peers and patrons. But the discipline of the postcard encourages, though by no means guarantees, the promotion only of those ideas which are clear, key and tried and tested. The postcard format lends itself to the kind of technologies which can be adopted on the basis of a single message – identified in Chapter 1 as robust technologies. The composting latrine was never made the subject of a postcard since it did not have that conceptual simplicity. In a booklet, ideas of marginal importance and questionable worth are liable to be thrown in as padding. The content of each postcard is a product in itself and has to stand or fall entirely on its own merits.

The Centro Sinchaguasin cards were seen primarily as a memory aid or souvenir for those people who had come to the Centre and seen the actual technologies. Cards produced by UNICEF to promote ORT in Nepal (see Figure 10) were one component in a broad campaign. Taken on its own, the drawing on the ORT cards might be criticized for being rather small and hard to understand; the dismembered hands and levitating tumblers would appear to be

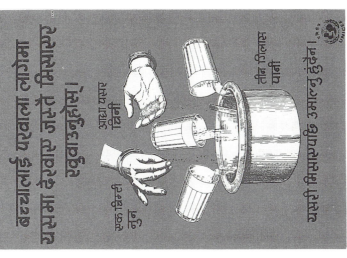

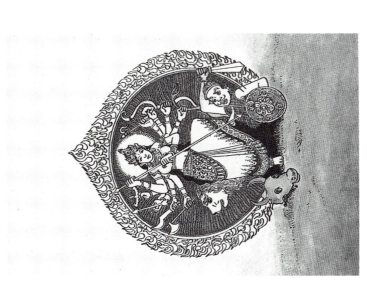

Figure 10 UNICEF health promotion card
A colour picture postcard promoting oral rehydration treatment in association with a local God in Nepal.

potential sources of confusion. However, the same drawing was used in a wide variety of contexts at several scales. It was used on posters and flip charts used by field workers explaining the technique of ORT directly to the villagers. The image was also printed as a decorative motif on fabric. The fabric was sold commercially and used to make up clothing as well as curtains for health centres. The drawing was always used in conjunction with a simple slogan, *Nun Chini Pani* (Salt Sugar Water). By widespread repetition, in conjunction with training, the drawing became not so much a means of conveying information as a prompt to trigger a memory of something which had already been learnt.

In any consideration of visual educational material it is normal to think in terms of printed matter. Studies of technical education programmes tend to focus on the booklets, posters and postcards because they are tangible and lend themselves to study. In practice, there is a mismatch between means of education and means of learning. A negligible portion of the typical villager's knowledge about technology has come about through graphic material. The overwhelming bulk of indigenous technical knowledge comes from observing and using the technology itself.

READING OBJECTS

Structuralist philosophers argue that although the objective world may consist of actual things, the world as we experience it is composed exclusively of signs. The mind has to decode the signs in order to construct a model of the world about it. Our thoughts about the world relate to the model in our minds rather than the actual world. In order to influence the mental model, it is necessary to manipulate the signs to which our senses are exposed. Semiotics, the study of signs, has generated a large vocabulary for describing the operation of signs. Charles Peirce, a pioneer of semiotics, identified sixty-six different types of sign. Fortunately, he classified these into three broad types: icons, indices and symbols (Peirce 1966: 391):

- **Icons** An icon is a sign which is determined by some actual physical characteristic of the entity which it signifies. Examples are a 'stick man' on a road sign and a graph which physically reflects the form of some abstract aspect of the actual entity.
- **Indices** An index is a symptom of something else. Spots on the

skin may be an index of some other malady and a point of reflected light on a sunny day may be construed to indicate a distant corrugated iron roof.

- **Symbols** A symbol comes to represent something else through convention and so has to be learnt and cannot be readily deduced. A rendered facade or a designer label is a symbol of prestige.

Symbols are culturally dependent and may be highly location-specific. In Ladakh, an eminent Buddhist scholar said that in Tibetan Buddhism there is an inviolable rule that monasteries have a red band painted around the top of the building, while houses have either a black or white band. In Central Ladakh, this was observed to be a consistent rule yet, in the neighbouring valley of Zanskar, also a Tibetan Buddhist area, all the houses had a red band.

Symbols are important in signifying value, but icons and indices are important in revealing the presence of specific entities which may in themselves be status symbols. The icon or index for the entity becomes a sign of quality. An index of quality may be the small window or slot which indicates 'the bathroom' inside, or the discontinuity in the bonding of a masonry wall which indicates the 'internal dividing wall' beyond. In Ecuador, a village built a large communal hall to a plan which I had given them. In that particular locality it was traditional, on tiled roofs, to paint the edge of the roof and its ridge with a line of limewash in order to ward off the evil eye. On this building, the villagers not only painted these lines but they painted two more lines which indicated where there were internal dividing walls below. Normal domestic buildings are not large enough to merit internal walls except maybe screens of wood and mats. The community's possessing a multi-roomed building was a sign of status which they wished to express.

An index may signify quality through indicating an entity which is not even physically there. A familiar sight in Third World countries is that of reinforcement steel protruding from the tops of houses. The steel is an index of the otherwise hidden reinforcement, but it can be seen also as an index of the aspirations of the owner for an extra storey.

In Yemen, the Dhamar building education project proposed the use of a reinforced concrete ring beam which could be added to an existing house. The ring beam was visible on the face of the house. Several house owners had decorated their ring beams. Such a concern to celebrate the presence of the ring beam suggests that the display value of the technology was considered important.

After the Ecuador earthquake in 1987 it was recommended to householders that a timber ring beam should be half-sunk into a channel cut into the top of the rammed-earth wall. The channel provided a better key between the roof structure and the wall and so helped the roof to stabilize the walls. In traditional roofs, light filters in under the eaves and between the rafters. Consequently, within the room there is a regular series of patches of light corresponding to the gaps in the rafters. In the new houses, the slight lowering of the roof reduced the intensity of the light patches. To villagers in the reconstruction area, the quality of light in the room became a widely understood index of whether or not the house owner had built his roof properly.

In the design of the 'L' mould for rammed earth the inside corner was chamfered at 45°. The chamfer acts like a gusset in order to reinforce the corner by avoiding the damaging concentration of forces which an earthquake brings to bear on a sharp 90° corner. Although the carpentry required to make the moulds was more complicated, it was considered worthwhile for the structural benefit achieved. In practice, the chamfer was found to have another unexpected and, arguably, greater value. In the finished house, the chamfer appeared as an angle in the corner of the room. This meant that any person entering the house who was initiated in the construction system could instantly recognize whether or not the house was 'well built'. The chamfer became an index of building quality. A couple of other reconstruction projects used 'L' moulds but without the chamfer. In their finished houses, it was only by careful inspection that one could determine whether the house had been built using the 'L' moulds or the conventional straight moulds. There is less incentive to adopt a progressive new technology unless one's peers can see that you have adopted it. The chamfer acts to increase both the legibility of the building fabric and the status of the house owner.

The above examples are all accidental instances of some aspect of a technology becoming an important index of the technology's presence. These experiences suggest that in the development of a technology the designer should be on the look-out for possibilities of features which could enhance its legibility, just as commercial designers try to visually emphasize a product's distinctive features. At one stage in the design of the 'L' mould, we had considered putting a small chamfer on the outer corner of the 'L' as well as the inner. In terms of seismic resistance this would have made no difference, it would simply have protected the exposed corner of the earth wall from accidental damage. In the end, we omitted this detail since it seemed an unnecessary complication. But, in retrospect we realized that the external chamfer could have had a symbolic importance quite disproportionate to the marginal cost of the additional triangle of wood. Whereas the inner chamfer could only be seen by visitors who went inside the house, the external chamfer would clearly express the use of the 'L' mould and hence the status of the owner to any passer-by who understood the code. In a subsequent project, when an 'L' mould was developed for use in Pakistan, the symbolic possibilities for the external chamfer had been recognized so both the internal and external chamfers were used.

ARTICULATE TECHNOLOGY

The use of an index, such as the chamfer, reveals the presence of the technology to people who know what to look for. In itself, such an index would not be enough to convey the essence of a technology to someone who was not already familiar with it. In a demonstration centre the task is different. The visitor is being asked to try and understand a new idea. It is not enough simply to signal the presence of a technology, it must help to explain itself. The more legible a technology is, the more it will assist the visitor to decode the technology in order to construct his mental model.

At Centro Sinchaguasin there was a demonstration house. The plan of the house can be described as two thick walls which wrap together to form two spaces (see Figure 11). One half of the larger of the two spaces is covered with a roof. A lightweight timber and glass wall separates the inside from the outside. The heavy wall which encloses both house and garden is the key entity. Indeed, the 'house' comes to include the garden. The house as designed reflected my education as a twentieth-century architect. The form of the plan owes much to the 'modern movement' ideas of courtyard houses in which the garden is seen as an extension of the house and vice versa. The lightweight dividing line between inside and outside is regarded, in architectural terms, as almost not there.

After we had shown people around the Centre, one of the principle objectives was to ascertain which features of the Centre

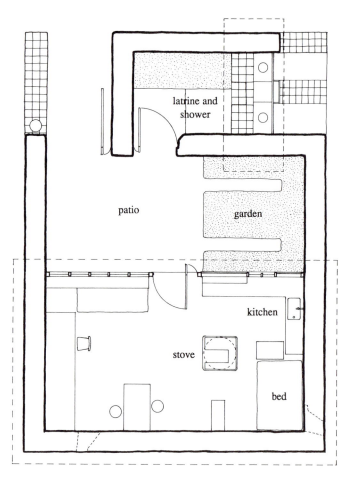

Figure 11 Measured house plan

had made the biggest impact. This was generally done through informal conversation. Some selected visitors were invited to try and draw a plan of our demonstration house from memory. Since the drawing of plans is not a universal skill, it was a task we could only ask certain visitors to attempt. The drawing exercise suggested that the villagers were breaking the demonstration house down into component entities which differed from those of architects. Three of the drawings are reproduced here; one by a village builder (see

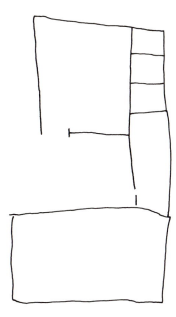

Figure 12   Plan by a builder

Figure 12), another by a child (see Figure 13) and the third by an Ecuadorian architect (see Figure 14). When possible, I watched villagers do the drawing and observed the sequence in which the parts were drawn:

- Invariably, the first thing to be drawn was the rectangle describing the room of the house.
- The second item would generally be the latrine enclosure. In the case of the plan drawn by the village builder, the latrine enclosure was drawn and then an eraser used to literally open the doorway linking it to the patio.
- The wall linking the house to the latrine would usually be drawn next. The opposite wall, next to the patio, often caused confusion. The builder left it out entirely, while even the architect was unsure where it ended. This may be because there was nothing – no entity – providing a conceptual landmark at the end of the wall.

The remaining sequence was less regular. The tanks of the composting latrine were always confused, and some people chose to

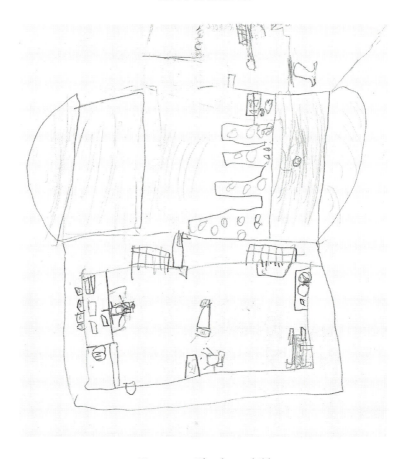

Figure 13 Plan by a child

show fittings such as the chimney and furniture. The main message was that the architectural entities of the two wrapped walls did not correspond with the perceived entities of non-designers. The formal conceptualization of house and garden as a single space, which is a deeply ingrained model for many architects, bears no relation to how most people, most of the time, think about buildings. Certainly in cold climates the most fundamental mental construct about the built environment is 'inside/outside'. However thinly the architect may draw the line on the plan, in practice, people are generally clear as to what constitutes 'inside'. The house form, as designed, was not consistent with an architecture composed of a

106

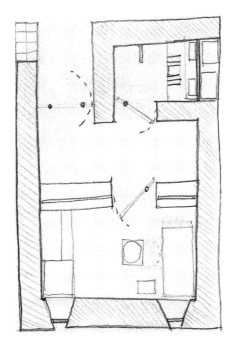

Figure 14 Plan by a local architect

series of intelligible, rational acts which a villager could imagine performing himself; such as:

- 'I'll build myself a house'.
- 'I'll build myself a latrine'.
- 'I'll create a patio by enclosing the gap between the two buildings with a pair of walls'.

The architecture of a demonstration house could be manipulated to stress the natural ordering of the component entities. The form of the house at Centro Sinchaguasin acted to obscure rather than to articulate this natural order. A house which was designed to stress the entities would become more house-like; that is to say, rather than the lightweight front timber wall it would have four walls all of the same material and it would have four distinct corners. The essential nature of the latrine is diluted by being integrated into the geometry of the architecturally determined walls. Because of its integration with the house, it had to be worked at in order to

107

understand its form. The latrine could have been more of a tangible, discrete thing, and the garden walls more like garden walls rather than a sculptural extension of the house walls. A second demonstration latrine was subsequently built adjacent to the house which derived some success from being a detached building. One could walk around it and take it in. For a technology to become articulate it needs to become more like itself.

THE PRINCIPLE OF CLEAR VISUAL MESSAGES

Recognize that different people see things in different ways. Before promoting a technology, assess every visual element for its ability to reinforce or obscure the essential message. In particular, look for opportunities which will allow the technology to announce itself in use.

Well-designed visual educational materials are no guarantee of successful development education. New ideas are usually acquired through observation, experiment and one-to-one dialogue with other people. While educational materials are often of little or no use in this process, they do have a limited place as tools to enhance the quality of educational experiences and as memory aids.

The Principle of Clear Visual Messages is not a simple recipe for success, rather it is a general plea for interveners to recognize that communication of technical messages is difficult. The intervener needs to ask of every educational tool, be it drawing, model, demonstration vegetable plot or building, 'who is it aimed at and how may it be misunderstood?'. The intervener should also ask of a technology, 'how will people be able to see that someone has used it?'. The social researcher will need to:

- Study the levels of literacy both of text and of graphic codes.
- Evaluate prototypes of educational material in the field to provide the designers with feedback on the efficacy of their material and the ways in which it is misunderstood.
- When technological introductions are being made, try and discover how they are being perceived and described by villagers. What visual aspects of the technology are sticking in people's minds?

The designer of educational material and of technologies has three tasks:

- To eliminate any unnecessary or potentially confusing elements in the design of any educational material.

- Preferably, to only use graphic codes which social research indicates will be understood, and where this really is not possible to signal the need for actively teaching the necessary codes.
- To look for opportunities to enhance visual features in the technology which may become the characteristic signs of that technology.

In Nepal, UNICEF have concentrated on promoting one exceedingly simple technical message, 'if your child has diarrhoea give him a mixture of salt, sugar and water'. They have recognized that this is a difficult task and applied a sophisticated approach using a variety of media. In technical education, the messages are often considerably more complex, and there are many of them. The greatest contribution to the legibility of educational materials will come from the policy makers in the development institutions cutting the number of key messages to an absolute minimum. The donors may best help by encouraging the production and use of effective material instead of rewarding institutions for the number and length of their publications.

Part III
RESPECTABLE

People want self-respect. Whatever their economic circumstances they still want to be respected by their peers. People only adopt ideas which they consider proper to people in their circumstances; suitable for 'people like us'. Increasingly, in a rapidly changing world this means being modern. An intervener must understand who or what it is that defines respectability. Often the criteria of interveners and the target population are at odds. The intervener must search for solutions which are acceptable to all concerned and, if need be, begin by example and persuasion to help redefine what is considered respectable.

Figure 15 Status display in a Ladakhi kitchen
In Ladakh, due to the cold nights and winters, the kitchen is the main living
room and is where guests are received. The kitchen shelves have come to
be the main vehicle for displaying status in the Ladakhi house.

6

MODERN IMAGERY

If man's need is what is necessary for his survival, his due is
what is essential for his honour – to his self-regard, and to the
regard in which he is held by others.

> (Michael Ignatieff 1984: 35)

The meeting of basic needs has been a recurring theme of development aid. In practice, it has proved difficult to establish a definitive
package which could reasonably be considered to constitute somebody's basic needs. Need implies necessity, yet it is clear that people
need very little merely to sustain the race. When a disastrous
situation arises, in which death is a real and imminent prospect,
the possibility exists that needs will become solely physical and
personal and that considerations of morality, self-respect and group
solidarity will be discarded. In Colin Turnbull's depressing account
of starvation among the Ik people of Uganda he describes the
complete breakdown of social values in which the survival of the
fittest became the only rule (Turnbull 1972). In such circumstances,
need has a real and more or less finite meaning but, despite
widespread suffering and poverty, the total disintegration of the
social fabric is rare. Generally, needs reflect a socially determined
minimum quality of life.

Needs are not absolute, nor are they static or universal. One
man's need is another's luxury. A person needs what he feels is
rightfully due to him in order to maintain his self-respect. Even
among the very poor, felt needs are frequently at odds with
externally defined and apparently 'objective' needs. For many, a
dignified death may be a greater need than a prolonged life. The
trade union movement in Britain was born in the burial clubs of
the eighteenth century. The need for a burial which could be seen

to be fitting and proper came before any organized concern over improved working conditions. As Peter Berger puts it: 'man does not live by bread alone – especially not in situations of desperate need and urgent hope' (Berger *et al.* 1974: 130).

In choosing between options of technology or custom, considerations of cost, practicality and utility are important, but the foremost concern, I suggest, is that of respectability: 'is this what people like us might do?'. If a potential option is not perceived to be respectable it will not achieve the status of an option. Without first being seen as an option, the other possible costs and benefits of a technology will not even be considered. The intervener needs to determine if and how a technology can come to be seen as respectable in terms of the intended beneficiary's hierarchy of values.

EMERGING VALUES

The massive scale and problems of rapid urbanization have been widely discussed in the literature on Third World development. Rather overlooked has been the even larger phenomenon which sociologists have dubbed the 'urbanization of consciousness' (Berger *et al.* 1974: 64). The impact of the modern urban life-style has been felt in the values of people who live far beyond the city limits.

Across the villages of the world, from cave dwellings in China to flimsy huts in Africa, there are walls decorated with pages carefully torn from colour magazines. Advertisements for Rolex watches cover cracks in mud walls. Posters and calendars depicting the Swiss Alps and the Canadian Rockies seem to have found their way into a large proportion of the world's roadside cafés. In Afghanistan, the carpet weavers have incorporated helicopters and anti-aircraft missiles into their patterns and their counterparts in Ecuador weave wall hangings of Elvis Presley. The Andean Indian has adopted the plastic gum-boot to accompany his other trademarks of the poncho and the hat; the digital watch is rapidly becoming part of the same uniform. In the industrialized world, appeals for famine relief are undermined by apparently incongruous television pictures of refugees bearing large radio-cassette recorders on their shoulders.

Meanwhile, among academics and professionals in the developed world, the value and appropriateness of traditional vernacular technology and design is being taught. It is argued that a solution which has evolved over the centuries will have achieved a good fit with people's lives and culture and is, therefore, appropriate. In practice, for many people who are consciously trying to assume an imported culture, that very fit may make traditional solutions unacceptable. The traditional solution may be economic, it may conform to the requirements of the villager's present way of life, but it may clash with the image of the way of life to which he aspires.

In 1981, a case study was published which described a typical household in Ladakh (Murdoch 1981: 261–78). The picture painted was of a traditional agricultural family working the surrounding land whose life was intimately intertwined with Tibetan Buddhist beliefs. Signs of change were noted, in particular two electric light bulbs powered from a communal generator and a metal stove which had replaced the former earthen hearth. In 1989, I sought out the same house in order to identify further signs of change (see Figure 2). Among the changes were:

- The father now works in paid employment as a labourer for the military.
- The son has grown up and trained as an electrical engineer and now works for the local radio station.
- Perhaps as a result of the son's training, the house has an extensive electrical installation including fluorescent tubes and a television with a voltage regulator.
- The floor of the main kitchen/living room has been covered with concrete, as has the stone stairway.
- Most of the cooking is now done with bottled gas.
- The downstairs 'winter kitchen' has been abandoned.
- A 'glass room' has been built on the roof with the intention of renting it to tourists in the summer.
- Whereas, before, people tended to sleep either in the kitchen or in rooms which doubled as store rooms, the teenage daughter now has her own bedroom with coloured walls decorated with posters.

Within a decade, this household experienced a range of significant changes all of which indicate a marked tendency towards a modern urban life-style. The 'glass rooms', described in Chapter 1, which were seen to be a widespread phenomenon across Ladakh may be just the most conspicuous indicator of a rapid and profound process of cultural urbanization.

West of Ladakh, in the remote valleys of the Karakorum mountains in northern Pakistan, a similar process of change can be seen. For most families the climate still necessitates living in the one room. However, there is now a high aspiration to build a guest house in which male guests may be entertained and visiting relatives accommodated. The typical guest house will have the principal characteristics of a 'down country' urban house: windows, a veranda, a smoothly rendered and painted front wall, a flat concrete floor, a corrugated iron roof, and, where it can be afforded, a pour-flush latrine. Despite the fact that these buildings are unusable in the winter, they are appearing all over the region.

In one village, in the harsh environment of upper Hunza, a large area of flat land has become available for agriculture because of a new major irrigation channel. The land is so attractive that the bulk of the villagers want to abandon their existing houses and build afresh on the new land. The villagers asked for assistance in producing plans of new houses. They were already quite clear as to the model of house which they required; it was to have three bedrooms, a living room, a kitchen and an integral bathroom – the urban model. When it was pointed out that the climate would still push them to sleep in the kitchen in winter, the villagers did not change their minds. The image dominated.

Both Ladakh and northern Pakistan had only limited contact with modernizing influences until the 1970s. At that time they became more strongly linked to the mainstream culture of their respective countries through the construction of roads. In the few years since the roads have opened the modernizing influences have rapidly taken hold. Elsewhere, obstacles like the Himalayas did not exist to obstruct the tide of change and so it advanced more rapidly.

In the 1950s, two authors – in separate studies from opposite sides of the world – highlighted the mismatch between an outsider's romantic view of traditional society and the insider's aspirations towards modernity. Daniel Lerner, in a study of cultural change in the newly oil-rich Arab nations, suggested that the people did not wish to be 'identified with what they were yesterday but rather with what they will be tomorrow' (Lerner 1958: 74). John F. C. Turner describes an experience in 1957 in which he used his best endeavour to design an attractive, low-cost school, using indigenous construction materials, for a village in Peru. The proposed technologies were rejected in favour of reinforced concrete since, 'for them the school meant a way out of an old and hopeless situation into a new and visibly hopeful one' (Turner and Fichter 1972: 135). Turner's experience, which has been replicated by many other practitioners, reveals that it is not simply that the villagers have lost interest in the old technology in the light of the new; they wish to positively exorcise the old technology and all that it represents.

At Centro Sinchaguasin, probably the biggest single mistake made by the Centre was in the design of the roof of the demonstration house. The house was roofed in thatch made from grass. The technology was entirely indigenous with no innovative changes. In the Ecuadorian Andes, with its combination of a high altitude and an equatorial climate, the difference between day-time and night-time temperatures is often great. A thatch roof with its naturally insulating qualities keeps the house relatively warm during the cold nights and avoids over-heating by day. It uses local materials and familiar skills. To the outsider it seems appropriate, but to the insider it is a symbol of poverty. The symbolism appears to carry more weight than any 'rational' considerations; or rather, the associations which the thatch roof carries make up a significant part of the rationale by which its rationality is judged. In terms of their criteria – their rationale – the rejection of the technology is perfectly logical.

Investment decisions of poor people can often appear perverse to the outsider. In a Latin American shanty town the visitor may be taken aback to discover that people may have extensive collections of smart clothes inside their makeshift shacks. This behaviour can be largely explained away by a sensible unwillingness to invest in houses built on land for which there is no security of tenure. But this is only part of the story. There is also the social need to express respectability in some form. If the medium of the house is denied them then people will find other forms of expression. If and when property titles are granted there is generally an immediate start on permanent building. Often the first thing to be done will be to build a facade onto the front of the shack. The facade of concrete block is usually rendered with cement and often decorated with bathroom tiles or some other form of ornamentation. Upgrading the rooms beyond the facade to make them more comfortable has a lower priority than establishing the public face.

Similar development decisions are seen increasingly among the rural population. A neighbour in Ecuador, Don Gonzalo, built his house facade long before installing electricity, piped water and elementary sanitation. The urban-style facade was followed by the even greater investment of laying a concrete floor surface in front of the house. The floor was surrounded by low concrete bench seats with decorative iron railings to the street. It became, as it was intended to be, the best dance floor in the neighbourhood. Not only did Don Gonzalo hold many parties but his friends and relatives borrowed the venue for their parties. His concrete patio was an object of pride and so a life-enhancing technology. Another neighbour living in a simple one-roomed house went heavily into debt in order to buy a large and expensive stereo system. While his wife and his mother thought this was foolish, it established his position among his peers as the man with the impressive stereo.

The process of judging our own and other people's households

Table 1 Check-list of needs, desires and dreams in an Ecuadorian suburb

	NEEDS *status-meeting*	*DESIRES* *status-enhancing*	*DREAMS* *status-changing*
Prosperity	*Low*	*Medium*	*High*
House	earth walls	cement block walls	two-storey cement block house
Finishes	untreated walls	plastered walls	decorated plastered walls
Roof	clay roof tiles of corrugated, galvanized iron	asbestos-cement sheets	concrete roof slab or glazed roof tiles
Patio	earth patio	paved patio	walled-in paved patio
Windows	window openings	glazed windows	large or arched windows with metal security grilles and curtains
Guest area	front porch	front parlour	a front parlour which is only used for guests
Bedrooms	bedroom/store	separate parental bedroom	separate bedrooms
Kitchen	kitchen with wood fire in special building	kitchen with water and gas cooker in special building	modern kitchen incorporated into the house
Water	outside tap	shower and water closet in special building	bathroom in the house plus water storage tank

is going on all the time. In the same way that the chamfered corner, described in Chapter 4, became a recognized sign of the well-built house, so too do other signs indicate whether one's neighbour is higher up or lower down the social scale than oneself. The process of evaluation can be thought of in terms of a three-column mental check-list of needs, desires and dreams. Centro Sinchaguasin was located on the outskirts of a small market town. The area was a rural suburb which was rapidly becoming urbanized both physically and in terms of consciousness. An exercise was conducted to identify the check-list being used by our neighbours (see Table 1).

In a study reported from rural Java, an analysis was conducted in which domestic features were also used as indicators for low-, medium-, and high-prosperity levels (Honadle 1979). Even though in substantially different cultural and ecological zones, the kinds of tendencies identified in both Java and Ecuador point towards the same urban model. The outsider may use such measures to indicate prosperity levels but to the insider they are also potent symbols of social value. In a rapidly changing context, like the suburb in Ecuador, people who have achieved the technologies in the medium-prosperity category would regard the lower prosperity technologies as not only representing poverty but as reminders of their own immediate past. To choose a technical solution in the low-prosperity list would be more than an economy, it would be an admission of social failure. While the high-prosperity list describes their dream house, the low prosperity list describes a nightmare to be avoided at any cost. One of the neighbours to Centro Sinchaguasin described the house reproduced in Figure 16 as 'the most beautiful house in town'. It embodies the urban concrete dream house. When I showed a photograph of the same house to people in Pakistan they thought it had been taken in Islamabad. The nature of the dream is becoming almost universal.

For most, the dream remains a dream. Table 2 shows how construction prices changed in an Ecuadorian village over a five-year period. The modern sector materials have moved further out of reach while the aspiration for them and the life which they represent has grown. As the realization of the dream becomes increasingly difficult, the need to express one's association with the dream is all the more urgent. Technological artefacts whose actual use would represent development to the outsider become symbols of advance to the villagers merely through their presence. A supply of electricity may, to an outsider, seem of primarily functional value

Figure 16 The universal modern house
This concrete house in a small market town in South America is instantly
recognized as the modern urban house by people all over the world. The
image is universal.

but it also represents status. In Ecuador, it is common to see a single
light bulb at the top of a pole in a patio burning all day to announce
to the world that the householder has electricity. The meter often
has pride of place next to the front door.

In a house in Nepal, the electricity meter was placed on the wall
over the fireplace in the living room. Electrical conduits radiated
from the meter in a meticulously executed variation on a traditional
Nepalese peacock motif. In Pakistan, where electricity wires tend
to be mounted on the surface of the walls, wires are laid side by
side and held down with wire clips. The clips are precisely placed
to make elaborate decorative patterns. In one house visited in
Pakistan, a light bulb and a light switch had been installed in the
main room even though electricity supply lines had not yet been
connected to that village.

Even where sufficient resources are available for the family to
purchase the material trappings of the dream, the exorcism of the

past is rarely complete. While people aspire to a city life-style, the bulk of their actual life-skills pertain to the rural past in which they were developed. Gas cookers stand in many old fashioned earth-floored kitchens where, in spite of decreasing fuel-wood supplies, they are seldom used. They are serving their symbolic purpose through their presence while, where possible, cooking continues in the familiar and understood fashion on an open fire. In northern Pakistan, possession of a water-flushed latrine is coming to be seen as a sign of respectability. Statistics from development institutions indicate that thousands of the ceramic pans necessary for 'pour-flush' latrines have been bought. On examination, there were many instances where the pan had not been installed. The pan sat in a corner of the veranda of the guest house, in some cases for years, as a statement about the nature of the way of life towards which the householder saw himself heading. In other cases, the pan had been installed but it was not used. The door to the latrine was locked, only to be opened for guests.

The peculiar phenomenon of people in northern Pakistan paying for a sanitation technology which they then do not use in part stems from unfamiliarity, but it is also a demonstration of a mismatch

Table 2 Comparison of construction prices, in Ecuadorian sucres, in an Ecuadorian village between 1983 and 1988

	1983	1988	% Increase
Sucre/US dollar exchange rate	70.00	450.00	543
Modern materials			
Wheelbarrow	1,750.00	10,950.00	626
50 Kgs cement	140.00	700.00	400
50 Kgs reinforcement steel	800.00	4,000.00	400
Lorry-load of washed sand	800.00	4,000.00	400
Ceramic W.C. basin	3,500.00	16,280.00	365
1lb of steel fixing wire	26.00	120.00	362
Concrete block	6.50	24.00	269
Indigenous materials			
2.5m rough sawn plank	28.00	100.00	257
Baked clay brick	5.10	18.00	253
Baked clay roof tile	2.60	8.00	208
Slaked lime	80.00	160.00	100
Labour			
Skilled builder per day	350.00	840.00	140
Official minimum monthly wage	8,500.00	14,500.00	71
Labourer	300.00	500.00	67

between the desired urban model and the practical demands of the rural reality. In many villages water is scarce and has to be fetched manually from distant sources. Often the house is built on solid rock, which makes it impossible to dig the necessary soak-away pit. In winter, where a latrine has been installed it is common for the water seal of the latrine to freeze. In one village, in the region of Hunza, the local field worker of a development institution had installed a pour-flush latrine because he believed it was the correct thing to do. After some discussion, he admitted that his family refused to use the latrine since they believed it to be a waste of valuable fertilizer. He was left in a state of confusion. He 'knew' that both sides of the argument were correct; all the evidence from his exposure to the modern sector told him that a flush latrine was what the modern household should have, while his whole history told him that the traditional practice made more sense. In other instances the conflict is not so black and white but none the less still present.

In Ecuador, while neighbours scorned our thatched roof, in more reflective moments one or other of them would admit that life under their old thatch had been much warmer than under their present clay tiles or corrugated iron roof. Similarly, the thin walls of concrete blocks are recognized as colder than the massive old walls of earth. Thatch roofs create objective problems such as dirt, vermin and the risk of fire. An earth wall is easily damaged and, if untreated, generates dirt. The undoubted advantages of the new materials in these respects are clung to as justifications for achieving the desired image while the increase of cold-related problems such as coughs and rheumatism is ignored.

PRESENTING A MODERN SOLUTION

To be considered an option a technology must appear respectable. If the indigenous choice process is coming to equate modernity with respectability then the challenge for the aid intervener is how to present his proposed alternatives as modern. One writer commenting on an aid programme to supply irrigation technologies in Bangladesh observed that 'the appearance of modernity was important . . ; to them it was synonymous with "the best" and the best was what in their view aid donors should provide' (Thomas 1975: 128). In the case of technical assistance, all the donor is providing is knowledge of an additional technical option. If that option is not considered modern it will be discarded. The more modern an option is, the greater the sacrifices which are often made in order to acquire it.

Of the technologies used at Centro Sinchaguasin the one which caused the most immediate excitement and enthusiasm among the neighbours was a roof made with a coating of sand, cement and sisal fibre over an insulating layer of earth, pumice and straw. The roof was finished with a coloured commercial water-proofing paint. Although the finishing paint was expensive, the base was so cheap that the whole roof worked out only slightly more expensive than corrugated iron while creating a better thermally insulated environment. The roof's smooth and coloured appearance made it look like painted concrete. A typical comment was 'it is better than Eternit'; Eternit being the trade name for asbestos-cement roofing sheets which were considerably more expensive and highly desirable. People would admire the roof, inside and out, feel its surface, and finally say 'it must be very expensive'. When told, at that stage, how inexpensive it actually was, the desire engendered was so great that it was sometimes necessary to persuade people *not* to copy the technology since, in functional terms, it was not reliable – it leaked.

The appearance of the roof was meeting the strongly felt need of indicating modernity. The decision that this was an interesting option came before any consideration of cost or utility. This is not to say that people would adopt a faulty and expensive technology simply because of its appearance, but rather that the image is considered first. Factors such as low cost and functional efficacy become confirmatory tests rather than the motivating force.

The task of presenting a product in a particular light is the task of advertising. Marketing and advertising are words which drive many aid workers to fury. The words conjure up images of exploitation and manipulation. But a body of knowledge, theory and insight has grown up around advertising, some of which, I suggest, may be legitimately applied to the problems of development communication. In the field of health care the concept of 'social marketing' is coming into common use (for example Birkinshaw 1989). The promotion of oral rehydration treatment in Nepal, discussed in the previous chapter, was reminiscent of a sophisticated commercial campaign with its co-ordinated use of a variety of media. Parallel to the campaign for home-made ORT was another for pre-mixed sachets of oral rehydration salts. A significant element in the attraction of the pre-mixed sachets seems to be their modern appearance. They are silver foil sachets labelled with coloured print which creates an image of high technology and modernity. The prestige of the product has been further enhanced by its promotion on posters with full colour photographs of equivalent quality to the ones used to promote prestigious commercial products.

A common concept among advertising theorists is that of the 'hierarchy of effects' in which the potential consumer is considered to move through a sequence of relationships with the product from ignorance of its existence to its purchase and use. The most commonly cited model for describing the hierarchy is the AIDA model which was proposed by E. K. Strong in 1924:

Awareness → Interest → Desire → Action

The hierarchy of effects models have been attacked since often two or more stages may occur virtually simultaneously. When someone buys on impulse, the more reflective stage of 'interest' may not really occur until after the purchase. In development aid, such impulse buying would appear both unlikely and undesirable. It may be assumed that some form of sequential progression is occurring. The AIDA model was designed to describe the purchase of finished products. A more recent model of the hierarchy, proposed by E. M. Rogers in 1962, attempts to describe the sequence with respect to the adoption of innovatory ideas rather than products:

Awareness → Interest → Evaluation → Trial stage → Adoption

This sequence provides a more useful description of a technical assistance process aimed at an independent self-determining villager. One consequence of considering the sequence is the realization that at each stage of the process the intervener may need to make different types of information available. If the initial 'awareness' information is of the wrong type – if first impressions are bad – the remaining elements of the sequence may never happen. If awareness and interest can be achieved then the villager is still entirely free to evaluate the technology fully and, if need be, reject it. At the awareness stage the villager does not need all the details. The intervener needs to feed the villager information which would lead the villager to conclude 'here is a product which it would be respectable for someone like me to have'.

The composting latrine attached to the demonstration house at Centro Sinchaguasin was unusual in that it took the form of a small walled garden in which the two latrine seats occupied one side. The seats were covered by a thatched roof, while the garden and a paved area with a shower were open to the sky. The design was executed in this way to challenge the image of a latrine as a dark, smelly little room. The latrine could be used in privacy and was sheltered from any wind while still enjoying sunshine, the sight of flowers and a vista of distant mountains. Foreign visitors and Ecuadorian intellectuals from the city loved it while local people thought it plain weird. In challenging the unpleasant image of a latrine, the image which replaced it was too unconventional.

The alien nature of the idea in which there is an ambiguous relationship between inside and outside was described earlier, as were the undesirable associations of a thatch roof. When combined with the novel idea of collecting and recycling human waste the whole effect was one which led to an early rejection. Since the latrine was in constant use over several years, immediate neighbours inevitably engaged, to some extent, with the stages of interest and evaluation. When the first fertilizer was produced and seen to be both rich and odour-free they expressed polite interest, but the first image had been wrong. For the one-off visitor the bizarre nature of the latrine tended to discourage any further interest.

When a second demonstration composting latrine was built, both the legibility and image of the technology were the paramount concerns. It was apparent that people wanted a bathroom to look like a bathroom. Since a bathroom is not a traditional rural feature, this meant a modern urban bathroom. The new latrine was a simple white box with a tiled roof. It contained a shower with a tiled tray and space for a wash-basin. In the first latrine, the woodwork of the door, seats, chimney and the like had been oiled to bring out the natural colours of the wood. In the second latrine, paint in pastel shades was used. By using different colours for different elements of the building the legibility was enhanced as well as the modernity of the image. The result of changing the image of the technology was immediate. After three years of indifference to a technology with an unacceptable image, the second latrine with the identical technology stimulated neighbours to ask about cost and other details while the last touches of paint were still being applied.

During the following three years approximately seventy latrines were built. The bulk of these were built with the assistance of the *Fundación Ecuatoriana del Habitat* (FUNHABIT). In one project, in the village of Cachi Alta, FUNHABIT provided latrine seats and chimneys while the villagers built the rest. FUNHABIT provided the seat already painted. They used a range of different colours to provide choice and to emphasize that the technology was not simply being presented for its functional worth. The latrine projects of FUNHABIT, although useful for promoting the idea, still involved a significant element of subsidy and control. Of more importance was the take-up by individual householders implementing the technology on their own initiative and with their own resources.

In Chapter 1, the example was given of a villager who had built his own single-tank version of the latrine. In another instance, a villager who was president of his community built a two-tank latrine which embodied the essential principles but which did not follow the precise details and dimensions of the demonstration latrine (see Figure 17). In both cases, while attention had been caught by the modern image of the demonstration latrine, once the technological principle had been adopted, the image of the implemented technology became less important. The tangible and desirable entity was no more than a vehicle for promoting the principle.

Figure 17   A self-built composting latrine

Moving through the hierarchy of effects, once awareness has been generated and interest aroused villagers may wish to conduct their own trials. This can involve the intervener with different forms of input. At this stage, in which the villager is actively seeking information for his own purposes, the booklet or the training course may be suitable for some technologies. Alternatively, access to physical demonstrations may be better. With the composting latrine it was useful to have two latrines, each conveying different types of information for different moments in the process. One was in operation in order to demonstrate the technical principle while the other was left unused so that visitors could climb into it and inspect the details of the construction. Neither was fully adequate on its own.

In some cases the intervener will need to provide a material input to enable the villager to conduct a trial. In the Yatenga water conservation project a length of transparent hose was required to make the 'water level'. At some future date when the use of such levels becomes commonplace the market might be expected to provide the hose. But in the first instance the project workers had to bring the hose from the capital city in order for the idea to be widely tested.

With new crops the aid worker may need to supply small samples of seeds in quantities which would not normally be available through the commercial market. Once a villager has tried the new seeds in a small corner of one of his own fields he may be prepared to invest in a commercially significant quantity. In some projects the intermediate stage between demonstration and full-scale implementation is disregarded. The farmer's interest may be aroused by a demonstration, but this does not mean he is prepared to take the risk of investing in something which he has not proved for himself on his own land.

A common task for the advertiser is to take an old and stale product and re-present it to appear new and exciting. Although this may seem a distasteful and exploitative activity, many Appropriate Technology projects are faced with precisely this challenge. All around the world, aid workers are trying to persuade people to plant native species, restore ancient irrigation systems and build with local materials. The old ways are presented as the wise ways. In Latin America it is common for aid workers to speak of *rescate cultural* – 'cultural rescue'. The aid worker will dress up in a poncho and speak of the wisdom of the Incas. Yet, given a choice between the irrigation techniques of the defeated Incas and that of the prosperous Californians, many villagers will show an understandable preference for the latter.

The selective sentiments of nostalgia seem to appeal more to the emotions of the European and North American urbanite and the urbanized Third World intellectuals than to the rural poor. For most people, the past is the problem not the solution. If some of the old ways do indeed offer objective benefits then means must be found to encourage people to use those techniques in such a way that they feel they are moving forward rather than backward.

In Chapter 4 it was suggested that an 'ingredient X' can offer an excuse for change. Paradoxically, the ingredient X can also provide a reason for carrying on much as before. Familiarity breeds contempt, and in the light of new alternatives a traditional solution may well seem undesirable. A commercial manufacturer with an established product will often make some minor alteration in order to present it as up to date and also as an excuse for restating the product's original virtues but in the context of a demonstrably new product. He achieves the best of both worlds; reassuring familiarity with a satisfying modernity.

The use of the 'L' mould in Pakistan provided the incentive for a reassessment of the value and potential of rammed earth. When building their new health centre the villagers took great care to build walls which were smooth and vertical. The result was a quality of construction far superior to traditional rammed earth construction. This improvement had little, directly, to do with the 'L' mould; it was simply a flagship technology which triggered an ambience of modernity and quality. In Ecuador, we treated the outside of earth walls with a limewash containing brightly coloured pigments. The result was a modern-looking wall with a clean surface which retained the advantages traditional earth walls have always had over concrete walls.

Normally, in the context of aspiration for concrete houses, people would forget about the benefits of traditional construction. They would almost conspire to let their old houses fall into disrepair as a means of justifying their aspiration. When the old technology was re-presented in a new and modern light the benefits of a relatively well insulated and low-cost wall would rise once more to the surface. Once people have latched onto an idea as an option they seek out confirmatory facts, some or all of which they already know but have suppressed. The ingredient X becomes a catalyst for reassessment.

RESPECTABLE NAMES

In Chapter 4, it was suggested that people tend to think of their environment in terms of the constituent entities. Consequently, problem statements will tend to be focused on entities. Each entity comes with a name and a set of associated values.

The people whom one associates with a technology make up a large portion of the values connected with it. If a technology is conventionally associated with a labourer it has less prestige than something used by a prince. In the next chapter, the influence of opinion-formers in society and their possible exploitation by aid interventions is discussed. But many technologies, and their names, are already firmly tied to certain levels in a hierarchy of social values. Names like window, door, concrete block, mud-brick, house, shack, toilet and latrine all carry values which may or may not coincide with the package of values associated with one's self-image and aspirations. A consideration of the meaning of names can lead to important questions for new technologies and interventions. For instance:

- When does an improved mud-brick become a low-cost concrete block?
- When does a shack become a house?
- What is the difference between a latrine and a toilet?

To an intervener the concept of an improved mud-brick may satisfy a number of criteria regarding use of local materials, energy conservation and indigenous development. To the villager the mud-brick, like the thatch roof, may represent the problem he is trying to solve. The villager's solution may be concrete block because that speaks of the city. In some cases, there is the possibility of an overlap in which the same physical entity, an earth-cement block, may satisfy both sets of values when presented in two different ways.

In the promotion of the composting latrine it had been intended to use the title *la letrina abonera* – the fertilizing latrine – but we questioned the perceived meaning of the word latrine. For the type of people who are likely to read this book, a latrine is not a thing they would expect to have in their own home. No property agent would cite a latrine as a selling point to a house. Yet, repeatedly, development programmes expect people to desire a latrine. As a general rule, a latrine is not something which people aspire to, rather it is something an organization prescribes for others. It is for them not us.

In Ecuador, it seemed that either people did not place any priority on sanitation or else they wanted a 'proper' water closet like those that people have in towns. Any notion of a gradual scale of increasingly sophisticated sanitation technologies was not reflected in actual practice. The latrine is an intermediate model which exists primarily in the minds of outsiders. The composting latrine was renamed the *inodoro abonero*, since *inodoro* is an understood and respectable word for toilet meaning, literally, odourless. The alternative expression, *servicio higiénico*, is taken to mean water closet; to have used that term would have been misleading. On a promotional poster the title was linked to a promotional slogan; *La solución moderna – el inodoro abonero*. Whereas a latrine might have been promoted as the appropriate thing to do, the *inodoro abonero* was presented as the fashionable and modern thing to do.

THE PRINCIPLE OF MODERN IMAGERY

Present one's proposed option as being respectable, which will increasingly mean presenting it as the modern thing to do.

The Principle of Modern Imagery, above all, obliges the intervener to regard the intended beneficiaries as human beings who behave in the same way as any other people – wealthy or poor. People need to feel self-respect and to feel that they are respected by their peers. A new technology is only likely to be adopted if it reinforces and enhances the user's self-respect. In a growing range of cases this means identifying solutions which are perceived as modern – not appropriate, indigenous, sustainable or even in the first instance low-cost, but modern. In order to implement this principle in the design of a technology, the social researcher needs to:

- Identify criteria by which villagers determine respectability.
- Determine what cultural connotations the various available indigenous technologies and materials may have.

For the technology designer, the challenge is:

- To devise technologies which are not only technically useful, but which also have an image that is demonstrably respectable.

7

INFLUENTIAL PEOPLE

America can be most significant for the rest of the world *by being itself*. Among other things, this means that America, the most modernized large-scale society, can be a vast laboratory for innovative experiments to solve the dilemmas of modernity.
(Peter Berger 1974: 15)

Modernity means choice, and choice is generally considered to be a good thing; but there is a price. Peter Berger describes the alienation caused by freedom of choice as the 'built in crisis of modernity' (Berger 1977: 134). This crisis is severe in the advanced industrial countries and may be more severe in societies which, in some cases, are being catapulted in to the twentieth century. Berger suggests that traditionally the strain of an excess of choice has been dealt with by the creation of institutions which protect the individual from having to make too many choices (Berger *et al.* 1974: 167). We choose largely to limit our own range of choice through the acceptance of social mechanisms which determine which choices are respectable.

In Chapter 2 it was suggested that development aid should be implemented through people and institutions that are recognized by the intended beneficiaries as competent in the relevant field. In any society there are also those who are seen as competent to place the stamp of respectability on a technology or practice. By introducing change through the accepted arbiters of convention, or 'change agents', interveners can help to ease the strain of rapid change.

TREND-SETTERS

In recent years, in Western societies, trend-setting has been associated with youth, and innovation and originality have been seen as trend-setting. But the people who are the first to use a technology are not necessarily those who establish a trend. With respect to technology change, it is possible to regard society as consisting of four groups (based on Haaland 1980: 10):

- **Innovators** Innovators are often younger people who are curious about new ideas and who can afford to fail. They offer the prospect of a gratifyingly receptive audience for the intervener's proposals. Projects which target single young men in jeans and sun-glasses may achieve rapid results, but their success may well be short-lived and limited in scope. Adventurous young innovators are often looked on by the rest of the community as gamblers whose opinions cannot be trusted or whose position as people without responsibilities is of little relevance to their own.

- **Respected minority** The real trend-setters are in this second group. The schoolteacher, the trusted shop-keeper, the farmer who can read or the craftsman who is respected for his work are people who can be open-minded, often have some education and whose opinions are sought. They will be the kind of people willing to take a calculated risk if the consequences of failure are not too great.

- **Majority** The bulk of the community is, generally, relatively conservative and only adopts a new idea when it has been convinced by the sight of the respected minority successfully applying it.

- **Laggards** The laggards are the most conservative people in the community and the last to change. Some people have religious and cultural reasons for being conservative, others are conservative because they are poor or because they are outsiders and anxious not to appear different. The poorest groups of people may be those most in need of an intervener's innovation yet, since they can least afford the risk of failure, they will be the most resistant to adopting unfamiliar and hence unreliable innovations. It is also unlikely that the poor will be emulated by the rest of society.

Membership of the respected minority depends on circumstances. Respect is not simply a consequence of wealth. A UNICEF study of a Nepalese town suggested that many small businessmen are among the respected minority, while more innovative big business-men are not, since they are regarded with suspicion. In Yemen, migrant labourers working on the modern sector building sites of

Saudi Arabia are a major source of knowledge and values. They return not only with money but with the mystique which accompanies knowledge of the outside world (Leslie 1987: 45). In lowland Ecuador, carpenters working in ship-yards in the nineteenth century returned to their villages with a knowledge of elaborate European wood joints. Such joints were incorporated into the indigenous house-building tradition and are still being used.

In the same way that the defence industry has been an important engine for the development of high technologies, the military has also been a major means for disseminating technologies and technological values. In eighteenth-century Russia, Peter the Great introduced a programme for training field barbers as medical auxiliaries. After they were demobilized the barbers returned to their rural communities where they continued to practice these skills (Hardiman 1986: 53).

The remote mountain region of Ladakh, in Indian-controlled Kashmir, is a strategically sensitive area because of its border with China and Pakistan-controlled Kashmir. When the military moved in force to Ladakh during the 1960s the agriculture of the region was insufficient to support the extra troops. Since the cost of importing food from the lowlands was prohibitive the army instigated programmes of research into high-altitude agriculture and subsequently initiated training programmes for local farmers. Self-interest on the part of the military led to significant improvements in local production and nutrition. The army also introduced the pressure cooker which, as mentioned earlier, has rapidly spread into domestic use across Ladakh. Similarly, in a study of improved cook-stoves in Kenya, Paul Harrison observed that returning army conscripts became an important route for disseminating knowledge of the new technology (Harrison 1987: 214).

In northern Pakistan there is a long tradition of army service. Over many years – long before formal development projects reached the region – young men left their villages to join the army, eventually returning with money and a new set of values, new knowledge and experience of organization. One army engineer had learnt how to make suspension bridges. When he returned to his village in the Hunza valley he realized that constructing just one bridge would give the occupants of the valley access to markets and services hitherto denied to them. He designed a bridge, calculated the costs, and then held meetings in the villages in the valley to present his scheme and collect money from every household. The

100-metre vehicular bridge was built and twenty-five years later it is still serving the valley. Arguably, this one bridge has made more difference to the development of the region than any other single initiative.

That the Aga Khan Rural Support Programme has been the catalyst for the formation of over a thousand village organizations in northern Pakistan has already been mentioned. Each village organization has an elected president and manager. No statistics exist, but it is evident that many presidents and managers are retired soldiers. Even where they are not, conversations suggest that there is widespread agreement among villagers and field workers that the retired soldiers are often the key activists. To some extent this is a product of their technical know-how, but it appears also to be connected with an attitude that sees change as both possible and within their power.

The task of promoting alternative low-cost building technologies among low-income groups in urban housing schemes is made difficult by two inter-connected factors; the people want that which is modern, and the organizations giving housing loans are generally only prepared to lend money for 'properly built' houses. When I was working with urban housing cooperatives in Ecuador in the early 1980s these problems actually became insurmountable. Some years later, a fashion grew up among the relatively wealthy urban intellectuals for earth-built houses which were considered 'green' and spiritually satisfying. This articulate group successfully campaigned to persuade a large credit cooperative to give loans on earthen buildings. A precedent was thus set for future finance and a growing fashion reinforced. By the end of the 1980s low-income housing cooperatives were coming to reappraise the alternative construction technologies. The intellectual middle classes had established a fashion which brought achievable low-cost solutions into the realm of what could be considered modern – and so respectable. Similarly, urban upper middle-class fashions for picturesque clay tile roofs are now making the rural population reconsider its recent rejection of such roofs in favour of flat, cold, leaking and expensive concrete roof slabs.

Berger suggests that 'the schoolteacher has been a carrier of urbanity for at least a couple of centuries'. Ivan Illich refers to the school as 'the new universal church' and Berger maintains the religion proclaimed in this church is 'the mystique of modernity' (Berger *et al.* 1974: 132). Conversations with staff, pupils and parents in various countries confirm the finding that education in a rural school is perceived, above all, as a passport to the city and the modern way of life (Hall 1986: 83). Several pupils in rural secondary schools in The Gambia stated that their objective was to pass exams so they could get jobs as waiters in the tourist hotels on the coast. More than one student went on to say this could lead to marrying a tourist and escaping the country altogether. 'Appropriate' education strategies, such as vocational training in crafts and agriculture, were seen by the intended beneficiaries as second-best options precisely because they were not urban-orientated.

As in the case of schools, there are differing perceptions regarding the message of rural development institutions. The local field worker is a new category of change agent and is in an ambiguous position since he or she is of both the community and the modernizing institution. In practice, he or she may be more of one or the other than is admitted. The educated local field worker in jeans may not any longer be truly in touch with his or her own community, and in some cases may be positively prejudiced against them – perceiving them as representing a life of ignorance and poverty which he or she has escaped. In other circumstances, the village-based employee of the aid institution may be as perplexed about his or her role as the rest of the villagers. Local field worker and villager are likely to hold a model of development in which the images of the city are the goal. Yet, with a crisis of modernity in the advanced industrial countries, the messages of modernity are growing confused and contradictory. The cities tell people 'concrete is best', while aid workers and urban intellectuals arrive in Toyotas, ponchos and sun-glasses to tell the villagers: 'build in earth, use medicinal herbs, save the forest'.

ROLE-MODELS

In highly traditional societies the role-models people use are likely to be local. In Nepal – in the course of designing visual images to promote health messages to rural women – UNICEF tested various pictures on a sample group. The pictures which depicted a city woman were unsuccessful because they were alien; the women could not relate to the people in the pictures. However, other pictures which had been carefully chosen to depict *typical* rural women were also unsuccessful. The all-too-realistic portrayal of rural poverty was not an image which encouraged emulation. The really successful image was that of a prosperous, smart and respectable member of the community. Unlike the picture of a city woman it was an image towards which the target audience could realistically aspire. She was a good example of 'one of us' (Haaland 1984).

In the modern world, the category 'one of us' has become increasingly hard to define. If the essence of modernity is choice, then a crucial arena of choice is deciding what sort of person one is. For many people – in the Third World as much as in the industrialized countries – the soap opera star, the TV anchor man and the newscaster have become not only part of the local community but a part of its respected minority.

Recognizing the increasing modernity and homogeneity of cultures, development projects have made attempts to adopt commercial advertising techniques. It is not uncommon to find posters for things such as vaccination which feature sports or television stars. Even among audiences which recognize this kind of figure, the impact can be less than was hoped for. Any attraction there may be lies in *emulating* a sports star rather than listening to what he is saying. If a famous sportsman in a suit is pictured with a child being vaccinated the connection will most probably seem nonsensical. A sportsman in his sports clothes drinking milk may be more effective, since the connection between milk, a strong man and health will be apparent. Despite a rapid urbanization of values and aspirations the process, in rural areas, is far from complete. For an audience who neither has television nor follows the sports pages in the national press, a sports star in a suit will clearly be meaningless.

The central message from commercial advertisers is 'sensible, modern people use our product, so if you want to be sensible and modern then you should use it too'. In contrast, development institutions are often in the position of implicitly saying 'I think this option, although quite unsuitable for a person of my elevated status, is suitable for you – do as I say not as I do'. And more often than not, people see through paternalism and resent it.

While development workers are promoting such things as mud bricks and pit-latrines in the villages they are seen to return to houses and offices built of concrete block, with water closets. Development institutions with their educated people, vehicles and concepts of democratization are prime agents of modernity. For villagers aspiring to modernity, the actions of development workers are more likely to be emulated than their suggestions. The development worker should be in a position of saying: 'Look, here's a really good idea. I think it's the best option available which is why I use it myself. If you like I'll show you how you can do it too'.

Wherever possible, development workers should seek opportunities to practice what they preach. In many circumstances it is not practical to see this principle through. A visiting expert from a European city cannot realistically live like a peasant farmer in a marginalized Third World region. But, Peter Berger's statement at the head of this chapter emphasizes that the task of helping people to attain their modern aspirations is matched by another equally important task: that of redefining what modernity looks like. The long-term aim must be for the European city and the Third World village to appear to exist on the same planet. They *already* share many of the same aspirations.

TARGETING THE OPINION MAKERS

The most effective interventions are likely to be those targeted at, or which come to be associated with, the respected minority. It may be hard to find strategies which can be so clearly targeted but examples do exist. In Nepal, UNICEF recognized the special status of the retiring Gurkha soldiers returning to their villages. The Indian and British governments were already organizing regular rehabilitation courses for retiring soldiers; UNICEF utilized this existing institution in order to train the soldiers in oral rehydration treatment (ORT). The benefits of this strategy were four-fold:

- For the soldiers, the training course in the context of military discipline was an understandable context. The institution of the military is seen as defining modernity. If ORT is presented by the military as the correct thing to do then it is simply accepted as the way things *ought* to be done.
- The notion that the retiring soldier is a responsible man who has something to offer his community is reinforced. Not only is ORT given credibility but also care for children's health is presented as a legitimate area of activity for the modern man.
- In the village, the retiring soldier is the archetypal arbiter of modernity. In village terms he is wealthy, he returns with modern clothes, he is associated with a major institution and tales of how things ought to be done. If ORT is part of the package of practices which he introduces then it is likely to be widely accepted.

- The geographical impact and cost-effectiveness of such a strategy is large. The returning Gurkha may live in a village ten days walk from the nearest road. The village may be beyond the reach of aid programmes yet, by using the existing agent of change, ORT may be successfully introduced.

Similarly, in India there are 800,000 soldiers retiring every year to all corners of the nation. Training of retirees is now being expanded to include issues such as forestry. But the use of the military is only suited to certain countries. Indeed, in Nepal some observers question how effective the promotion of techniques through retiring Gurkhas really is; it may be that they are so modern that they are too alienated from their own culture to be acceptable role-models. In other circumstances, there may not be a tradition of respectable military service. In Ecuador, the young returning military conscripts do not have the same status as the mature retirees of India, Pakistan and Nepal.

One of the most promising lines for future development research is the study of the agents of technological change and mechanisms by which they may be targeted. Some potential lines of enquiry are:

- **Health services** It was described earlier how, in northern Pakistan, interviews proceeded more successfully when carried out in the company of a health worker. The health services have an understood role which provides a point of entry for technological introductions. In particular this would appear to provide a natural route for environmental health improvements including sanitation and improved stoves. On the other hand, experience with Primary Health Care programmes suggests that the clearly defined role of health workers acts as an obstacle when they try to change their work to encompass preventive health care. The doctor is still meant to be someone in a white coat who gives pills and injections. There is often a gap between the self-image of Primary Health Care workers and their image as perceived by the villagers.

- **Schools** Similarly, the rhetoric of education policy in Third World countries increasingly emphasises technical and vocational training orientated towards the 'needs' of a rural community. Meanwhile the villagers continue to see the school as a gateway to the city. The school has the potential to endow ideas with the aura of modernity yet its vocational courses largely fail to do so.

- **Television** In Ecuador, as in many developing countries, television

138

is widespread in the cities. In a rural but urbanizing neighbourhood such as that where Centro Sinchaguasin was located,
television is in every home which has electricity. In remote
corners of the mountains of Nepal, Ladakh, and northern
Pakistan satellite dishes may be found tuned in to Bombay soap
operas.

In the Dhamar reconstruction project in Yemen a video was
prepared which presented building techniques in the context of
a situation comedy. This was reported to be one of their most
successful education techniques. In Brazil, one of the most
popular television soap operas is *Pantanal*. It is set in the swamp
lands on the fringes of the Amazon jungle. Its success may be
largely attributable to the frequent appearance of scantily clad
young people swimming in the river, but it is proving to be an
effective means of promoting 'green' issues in Brazil. In post-war
Britain, the radio soap opera *The Archers* was designed to
promote agricultural and nutritional information. Today, pressure
groups lobby the producers of soap operas to include their pet
issues in the programmes because they know it will raise public
awareness more successfully than any other method. There
would appear to be the potential of developing the soap opera as
a development tool. Similarly, if the TV anchor man were seen
night after night sitting in a nicely presented earth house rather
than a steel and vinyl studio it could be expected that values and
associations surrounding these materials would rapidly change.

- **Tourism**   Tourism is a mixed blessing with economic benefits
often having to be set against social problems. But, for better or
worse, in many places tourism has been a major mechanism for
exposing people to modern values and tastes in clothes, food,
accommodation and social behaviour. If tourists are examples of
all that is modern then opportunities should exist for using tourist
facilities to promote a new type of modernity to both tourists
and local people.

- **Religion**   There is a long tradition in many cultures of ascribing
religious connotations to day-to-day behaviour which is considered good practice. In the Christian tradition the expression
'cleanliness is next to Godliness' was used to encourage hygiene,
while in Islam there are prescribed cleansing rituals. The village
holy man may not be a respected authority on how to cure
sickness, plant healthy crops or build a better latrine, but as a
respected and dignified member of the community he can help

139

to endow new technological introductions with respectability by using them himself.

- **The wealthy**   If poor people aspire to that which is modern then part of the task is to help the better off trend-setters redefine the nature of modernity. Trickle-down economic development may be a rather dubious economic philosophy, but the trickle-down definition of tastes and aspirations is clearly a potent force.

It should be remembered that the real objective is to benefit those most in need. There may be some technologies, which can be successfully targeted at the respected minority, which are not relevant to the poor. Priority should be given to technologies which the respected minority will accept and which the poorest people can still achieve.

THE PRINCIPLE OF INFLUENTIAL PEOPLE

Seek out ways of targeting one's interventions at the respected minority, particularly at those people, including development workers, who are considered competent to define the nature of modernity.

This is probably the most important of the eight principles. However sensible and well thought-out an idea may seem to an outside intervener the real test is whether that idea will be seen as respectable by the intended beneficiaries and their peers. The attitudes and perceptions of a society towards an idea are largely determined by a relatively small set of locally influential people.

8

MULTIPLE AGENDAS

Phrases which divide the world into the 'haves' and 'have-nots' overvalue bread and plumbing and devalue music and architecture. . . . Phrases like 'under-developed', 'backward', 'simple' – to the extent that they cover a whole culture – are equally defeating. If, instead, we draw on an image in which two adults – one experienced in one skill, another in a different skill – pool their knowledge so that each can use the skill of the other for a particular task, as when a foreign explorer and local guide venture together into a forest, much more viable relationships can be set up.

(Margaret Mead 1955: 299)

The purpose of development aid is to help solve problems. Less clear is *whose* problems are to be solved. The sincerity of economic aid has been widely questioned by critics who argue that such aid is designed for the sole purpose of solving the economic problems of the wealthy industrialized countries (for example, Hayter 1981, Linear 1985). Village-level technical aid has, on the whole, been regarded as more innocuous. Despite the opinion of some critics that Appropriate Technology strategies reinforce an inequitable status quo, considerations of vested interests and conspiracy theories have seemed less relevant. Generally, the technological intervener has been characterized as having a non-controversial, apolitical, philanthropic role in assisting the poor to overcome their own problems. In *Small is Beautiful*, Ernst Schumacher reinforces the impression that the task of technical aid is a straightforward one:

Poor people have relatively simple needs, and it is primarily with regard to their basic requirements and activities that they want assistance. But their own methods are all too frequently too primitive, too inefficient and ineffective, these methods require upgrading by the input of new knowledge, new to them but not altogether new to everybody. . . . Because the needs of poor people are relatively simple, the range of studies to be undertaken is fairly limited. It is a perfectly manageable task to tackle systematically. . . . If the job is, for instance, to assemble an effective guide to methods and materials for low-cost building in tropical countries, and with the aid of such a guide, to train local builders in developing countries in the appropriate technologies and methodologies, there is no doubt we can do this, or – to say the least – that we can immediately take the steps which will enable us to do this in two or three years' time.

(Schumacher 1973: 194–195)

This view, in such an influential book, is startling in its confidence. It demonstrates a total faith in the power of the technical fix and the ability of outsiders to identify clear-cut problems and introduce definitively correct solutions. The underlying attitude seems little different to that which governed the Green Revolution and other technocratic aid programmes of the 1960s. The difference was merely one of scale. Schumacher originally described Intermediate Technology in precisely these terms. His thesis was an economic one in which the most cost-effective workplace had an initial set-up cost somewhere between that of an artisanal workshop and that of a modern factory (Schumacher 1973: 175–7). Yet, of the nineteen essays contained in *Small is Beautiful*, only one deals with this purely economic argument. It was apparent from the outset that the concept of Appropriate Technology was always intended to encompass something more.

MUTUALLY APPROPRIATE TECHNOLOGY

Numerous attempts have been made to try and define what 'appropriateness' in technology actually means. It has proven to be a far from trivial task. After a generation of promoting Appropriate Technologies no single widely-accepted definition has emerged. In 1980, Witold Rybczynski identified thirteen different published definitions of what made a technology appropriate (Rybczynski 1980: 1).

While certain desirable characteristics, such as the use of

indigenous materials and production in small local workshops, have tended to appear on most of the lists of criteria, the central emphasis of the definitions has varied considerably. Some laid stress on economics, some on conservation of non-renewable resources, and others on a technology's apparent capacity to promote communal activities. The common thread running through such definitions lay in their manifesto-like quality. Each was the expression of an intervener's ideological perspective rather than a description of how a villager determined the suitability of an option for himself. In a more recent and exhaustive review of the Appropriate Technology movement, the definition of Appropriate Technology was declared to be 'a technology tailored to fit the psychosocial and biophysical context prevailing in a particular location and period' (Willoughby 1990: 15). Such an all-embracing definition may be true, but it is useful only in that it reveals the futility of attempting a definition. A technology's appropriateness depends on the criteria of the observer.

In Ecuador, through a process of indigenous innovation, a new technology evolved among family workshops of brick and tile manufacturers in the district of La Victoria. The technology was a small mill which could be driven by hand or by motor to break up materials from which a glaze could be made. The glaze could be fired at a low temperature, conserving fuel, and it could be applied to normally semi-porous clay roofing tiles to make them impermeable and of a pleasing greeny-brown appearance. The tiles found a ready market with the middle classes among whom traditional-style pitch roofs were becoming fashionable, but who also wanted good-quality, rain-proof roofs. Formerly impoverished families were soon living in large modern houses and prospering remarkably. Added to these other advantages the material for making the glaze was a readily available waste product. Such a technology comfortably satisfied many criteria for appropriateness and rapidly spread across the district.

Unfortunately, the material from which the glaze was made consisted of old lead-acid car batteries. When they were ground up in the mill the air became dense with the lead dust and acid fumes. Levels of infant mortality rose to the highest in the country, as did incidences of cretinism. Workers from various institutions tried to 'educate' the families about the dangers of lead and tried to persuade them to drop the practice. But the families were not ignorant, they realized the dangers and had struck their own balance between costs and benefits. They continued with the technology because it fulfilled their agenda better than the alternative course of continuing in poverty with no prospect of escape. Who is to say what is 'appropriate'?

In the field of Appropriate Technology for building construction, the technology which has probably received the greatest attention is compressed earth-cement blocks for walls, such as those used in the Zambian project mentioned in Chapter 1. Various research groups around the world have come up with their own versions of hand-operated block presses; one of the earliest and most famous being the CINVA ram from Columbia. Although these have enjoyed some popularity among aid organizations and some success among urban housing cooperatives they have not really taken off on the open market. To the interveners, particularly the engineers, the block press seems an attractive solution. With the use of low-cost materials and unskilled labour a building block can be produced which is considerably stronger and more resilient than simple earth blocks. A mixture of earth and cement is placed in a metal mould and compressed by means of a lever to approximately half its original volume. The lever is then released and pulled back over the mould in a way which causes the block to pop out of the mould.

The numerous papers which have been produced on these blocks are characterized by graphs displaying the variations of strength which result from varying the percentage of cement, the degrees of compression, water content and the like. The exercise is seen as a technical one intended, by whatever means, to achieve the greatest strength for the least material cost. The villager, through exercising different criteria, is focusing on different problems:

- **Volume loss** From the engineer's perspective, the greater the compression achieved, the greater the quality and so the better the result; from the villager's perspective, blocks being compressed to half their volume means making twice as many blocks as before.
- **Slowness** Since several different operations are involved in making each block, the time necessary is greater than that for using simple moulds. In practice, some types of block press do not attain the desired degree of compression. In order to achieve the intended result the earth and cement mix has to be manually pre-compacted with the back of a spade. This slows up the operation still further.

144

- **Heaviness** In an urban situation the mould can be permanently based in a workshop and the blocks delivered to the building site. In rural areas, it is often more convenient, or even essential, to make the blocks on site. If the heavy metal press has to be carried manually or by mule, the option can become unattractive.

- **Robustness** The forces induced in the structure of a block press can be considerable. If the welds are of poor quality – as they often are – the structure can break and repairs are liable to be inadequate.

- **Ease of manufacture** Block strength is dependent on compression, which in turn depends on the operator pulling the lever down to its full extent. This is hard work. Towards the end of the day it is not surprising if workers ease off and quality deteriorates. Unlike conventional block making, the technology effectively excludes the use of many women and children in parts of the process. Many block-making workshops are family concerns.

- **Costliness** Block presses are promoted as being of low cost and, compared to motorized presses, they are. However, the capital outlay for a press, which has a risk factor since it might break, is considerably greater than for a wooden mould for mud bricks or even a simple metal mould for concrete blocks. Whereas every family can own a mud brick mould, a block press needs to be owned by an entrepreneur or a community organization.

- **Prestige** The kind of householder who is going to go to all the trouble of getting compressed earth-cement blocks is liable to ask: 'Why should I accept second best? For just a little more money I could have the real thing – concrete blocks.'

Engineers have tried to tackle the problem of poor compression. One of the more recent designs requires the operator to pump the handle of an integral hydraulic jack a couple of times to get maximum compression after pulling the lever. The laboratory experiments show remarkable results but, in practice, not only do the sophistication and cost of the mould increase, but the production of blocks is slowed down even more and the possibilities for error multiplied. The engineers are energetically solving the wrong problems. The criteria of engineer and villager are mis-matched.

Quite apart from conflicting perspectives between villager and interveners, there may also be conflicts of criteria within the aid community. In some cases, various interveners with distinct objectives and practices are working in the same villages. In Ecuador, after the earthquake in 1987, the affected villages were inundated by representatives of development institutions each of which had their own objectives resulting in a number of diverse 'appropriate' building technology packages. Some programmes aimed for maximum seismic resistance, others for lowest cost and greatest scope for replication, another for speed of recovery to allow people to return to economically productive activity, while an evangelical organization offered small numbers of prestigious houses in an attempt to purchase a foothold in the region. As soon as the villagers learned the game, their most appropriate course of action became whichever maximized their material benefit.

Three years after the earthquake, in one village with less than three hundred houses there were five different reconstruction programmes working side by side, offering packages which ranged from self-build mud brick construction to contractor-built prefabricated steel frames imported from Japan. Some families successfully exploited the programmes in order to obtain two or more new houses (Dudley 1992). Although a few families benefited greatly, any concept of appropriateness conveyed by any one of the intervening institutions must have been lost in the cacophony of conflicting messages. Such conflicts of criteria are inevitable in a pluralistic society. If these are not to be translated into contradictory actions on the part of the interveners, the existence of multiple agendas needs to be recognized and areas of overlap identified.

Even when a single intervener has no competitors, a self-imposed check-list of ideologically acceptable attributes for a technology can become a straight-jacket. Some of the technologies which fail the published tests of appropriateness have proved to be of immense value to the Third World poor. Smallpox vaccine is a product of the laboratories of multinational pharmaceutical companies but its important contribution to human well-being is indisputable. Flexible black plastic water pipe is a high-technology spin-off from the petro-chemical industry and is made in large and sophisticated urban factories. The benefit brought to villages around the world by plastic water pipes is almost certainly greater than that derived from all the technologies labelled 'appropriate' put together. In particular, the transformation in the lives of millions of rural women resulting from improved water supplies dwarfs the achievements of those development initiatives specifically pigeon-holed as 'women's projects'. In any commonsense assessment, vaccines and plastic water pipes are manifestly appropriate. Conversely, many technologies which have been consciously designed to pass the tests of appropriateness have failed to be adopted in the villages for which they were intended.

Plastic water pipe unquestionably has a potential for good. In fact it has several different potentials for different good outcomes. The potential which will be realized, if any, depends on the perceived need. In northern Pakistan, UNICEF have supported a large number of village piped-water schemes. In many situations, prior to the installation of the pipe, women had to walk long distances to collect water and the pipe successfully overcomes this problem. In other instances, the problem was not one of access to water so much as its quality.

In Chapter 3 the example was given of drinking water being taken from contaminated irrigation channels in preference to piped spring water. For the villagers, the key indicators of a water supply's cleanliness were that it was flowing and cold. The irrigation water came from glacial melt-water and so was colder than the relatively warm spring water. The irrigation water was constantly flowing whilst the tap water was self-evidently standing in the pipe until the tap was opened. To the interveners the warmth of the tap water was a positive attribute since it would remain flowing throughout the cold winter. This characteristic of functioning during the winter was also seen as valuable by the villagers, but the tap water was regarded as the second-best they fell back on when their 'good' channel water was not available. Both villagers and interveners were operating to two quite different agendas. In another, higher altitude community, the piped-water supply also froze during the winter so it was providing no perceived benefit. The villagers thus had no incentive to maintain the system which soon silted up and was abandoned. None of its possible potentials for good were realized.

By chance, in the first of the above cases of water-pipe installation, both intervener and villagers were satisfied. The intervener could tick off another name on the list of villages served while the villagers had a new and more convenient winter-time water supply. However, the intervener's main purpose for installing a water supply was to help reduce cases of diarrhoea. Since diarrhoea occurs principally in the summer the new water supply will have little effect as tap water seems to be little used in the summer. Rather like the solar-collecting Trombe walls discussed in Chapter 1, the result of the intervention is not a complete failure but it is different to what was intended. But since the villagers were clearly happy to have the water supply, this discrepancy between intention and result would not automatically come to light in the form of a clear rejection. The interveners should have carried out more sophisticated tests to determine whether their actual criteria were being satisfied. It had not occurred to them that an apparently straightforward technology like a water pipe could be valued in two distinct ways.

An intervener's criteria of appropriateness will only prove relevant if the intervention process generates options which also satisfy the criteria of the villagers. It is not necessary for both parties to share all of the criteria. Indeed, it is not necessary for them to share any criteria. The Big Idea of a single technology is frequently different for different people. All that matters is that a single technological solution can be simultaneously perceived as satisfying both parties. Once the intervener has accepted the possibility of dual agendas then the potential exists for the intervener to meet his objectives indirectly. The intervener may have to get involved with issues which are not directly related to his own objectives but which, from the villager's perspective, are a necessary part of a total package.

At Centro Sinchaguasin, the image of modernity, discussed in Chapter 6, was recognized as an important element in the villager's agenda. In the case of the composting latrine, the strategy was employed of showing how a modern-style bathroom could be achieved while at the same time we could meet our own objective of improving standards of environmental health. An alternative strategy might have been to promote the agricultural and hence economic benefits of the system and still achieve the health objective. An insulated floor was proposed primarily for reasons of improved health through the provision of a warmer surface for those who have to sit or sleep on the floor. In practice, greater interest was expressed over the more readily apparent benefit of a surface which was more cleanable than traditional earth floors. It was possible to meet both priorities with the one product.

In Nepal, the Annapurna Conservation Project is run by the Government in collaboration with the World Wide Fund for Nature. The project designers realized there were two perspectives: that of the project to combat ecological degradation and that of the local people who wanted to make money out of tourism.

Rather than choosing conventional, counter-development, policing strategies, the project presented itself as a source of support and encouragement to the local hoteliers and entrepreneurs. The project ran courses on how to cook food which the tourists would like, how to provide more acceptable dormitory accommodation, how to do book-keeping and marketing and how to speak English. Among the package of measures, the idea was promoted that tourists came in order to see natural beauty and that it was in the long-term economic interest of the hoteliers to conserve that beauty. The villagers now work in the off-season to clear up litter, plant trees and repair paths.

While in some cases the only people involved are an intervener and a villager, there are usually a large cast of actors such as a donor, the local development institution, government officials, manufacturers and men and women, all of whom have to perceive the solution as meeting their unique criteria. Some technologies, such as a steel stove project in Ecuador, satisfy the ideological criteria of the intervener and are easy and cheap to make for the manufacturer, yet the villagers simply do not like the product; the stove can only be used with pots of a certain dimension and it requires dry firewood of a certain size. Fibre-cement roofing tiles may meet the intervener's criteria for local production, relatively low cost and largely local materials; the villager, when shown sample roofs, is enthusiastic; but in the practice, the quality control necessary for production does not correspond to the day-to-day reality of small village workshops. A pour-flush latrine may meet an intervener's objective for hygienic sanitation, it may satisfy a man's desire to demonstrate the modernity and prestige of his household, but the person responsible for collecting water from a distant source may regard it as an unacceptably repressive technology.

Where it is possible to achieve mutually satisfactory perceptions of the same technology from different standpoints, the intervener may proceed happy in the knowledge that he is not imposing ideas but simply choosing to stress certain positive points more than others. But the useful new technology which succeeds in combining a fortuitous coincidence of interests is not always achievable. Once the existence of multiple agendas has been accepted then the possibility exists of situations in which solutions which are mutually satisfactory cannot be found. In such circumstances, one can no longer evade the question of whether mutual satisfaction is the only criterion. Can it be legitimate to address issues which are not felt by the intended 'beneficiaries' as their needs?

A RIGHTEOUS IMPERIALISM?

It is both easy and legitimate to describe the history of western imperialism in terms of a ruthless capitalist exploitation. Yet, even the most hard-nosed Marxist critic could hardly deny that many of the individual Europeans who went to the colonies were motivated by a genuine conviction that they were doing good. Many missionaries and colonial administrators saw their role as that of teachers bringing enlightenment to the ignorant. Whether the knowledge was of the Christian God or of how to make the trains run on time, the task was seen as one of education. European knowledge and values were sincerely believed by well-intentioned people to be definitively correct.

In more recent years the pendulum has swung the other way. Bottom-up development, enablement, empowerment, addressing felt needs and listening to 'the people' are among the popular slogans of development aid. The progressive aid worker is characterized as a dynamic 'Mister fix-it' who enables the community to realize its own agenda. The 'do-gooder' is now a figure for ridicule if not contempt. The criticism of old-fashioned paternalistic aid is summed up by Teresa Hayter:

> The assumption that foreign experts know best has ruled for too long, has racist overtones, and needs to be challenged; the capacities of ordinary people throughout the world, to make their own decisions and to determine their own priorities have been systematically under-rated and under-valued.
>
> (Hayter 1987: 104)

Today, for the aid intervener to be seen as in any way imposing his or her moral or cultural values on to other people is simply not done. In public pronouncements, the word of the community is held to be sacrosanct. If it were taken literally, this stance might readily be shown to be absurd. If we place a hypothetical and ideologically sound aid worker in a village, current wisdom requires him to say something along the lines of: 'I am here to help you, to listen to the needs which you identify and to try and work with you to develop simple tools which may help you in your tasks.' The reply could be along these lines: 'Thank you, that is very generous. Since we are cannibals, what we really need is an improved man-trap which can catch two or even three people at once. Could you help us with this?'. The professed philosophy suggests that the aid worker should then earnestly jot down careful notes about the community's needs and, in a suitably participatory manner, set about the task of devising some Appropriate Technology. In practice, most of us would applaud the aid worker for replying in a different manner: 'No. Not only will I not help you, I shall try and persuade you to change your current practice and if persuasion is inadequate I shall use all the power I can muster to compel you to stop.' The aid worker is making a moral choice; he is deciding that the principle 'eating people is wrong' takes precedence over the principle of unfettered self-determination. Peter Berger, discussing a similar example, points out that there is a difference between respecting and accepting the values of others:

> To understand is *not* to choose, but to accept as facts the choices of others. To act politically *is* to choose – and that means to choose between moral alternatives. Such choice, when there is power behind it, inevitably means imposing some of one's values upon others.
>
> (Berger 1974: 148)

The cannibal example is extreme, but the moral dilemma which it provokes is reflected in many day-to-day examples of development aid. Television documentaries portray indigenous jungle communities, perfectly attuned to nature, being eradicated and forced back by colonists. This is certainly true, but it is not the whole truth. In a meeting which I observed between a community of indigenous Amazon people and a representative of a donor agency the community was quite clear about its perceived needs. They wanted financial assistance for hiring bulldozers and chain saws to clear a large patch of jungle so that they could establish a cattle ranch and plant oil-palms. They did not want to conserve their unique way of life. They wanted a piece of the action.

Elsewhere, it is not unusual for communities, in the first drafts of their project proposals, to ask for funds to buy shotguns to protect their assets from thieving neighbours. In Ecuador, in projects to build community halls, the villagers invariably expressed a preference for brick or concrete over traditional earthen construction. The response of the donors to such requests was generally a discreet 'no'. The donor would let it be known that in his opinion the environment should be conserved, that shotguns are not nice and that indigenous building materials are good.

The interveners use various mechanisms to quietly exercise

criteria which are external and unrelated to the expressed needs of the community. The interveners may listen to the people but only choose to support those who are heard to say the right thing. The intervener as 'enabler' will only enable those actions which he wishes to support. In order to benefit, the intended beneficiaries must play the game as defined by the intervener. Even when the intervener sincerely wishes to respond to the community's expressed needs he can never completely escape his own responsibility of choice. To start with, the intervener has the task of determining who 'the community' actually is. As Margaret Hardiman observes: 'It is easy enough to go into an African village, have long and seemingly fruitful talks with the chiefs and elders, without any idea as to how far their views represent the interests of the people as a whole.' (Hardiman 1986: 57). The village community may contain fundamental injustices: women may be oppressed, the handicapped may be rejected, a caste or an ethnic minority may be disenfranchised. The intervener may choose to listen to the established leaders or to representatives of one of the oppressed sub-groups of the community. Either course involves a moral and political choice which cannot be evaded.

Hidden agenda have their roots in a variety of ideologies. The Victorian missionary quite unashamedly used aid as a lure to converts. Some of the evangelical North American donor agencies are continuing in this tradition. Third World political parties use aid programmes as a means of establishing a power base. This can be interpreted as a cynical practice to buy votes or as boldly putting words into action depending on one's perspective. Foreign governments, whether capitalist or Marxist, have used aid to promote their view of how the world ought to be.

Harder to pigeon-hole are the donor agencies and individual aid workers who are using the aid process in order to exorcise their discontent with the modern world. For these people a good project is one in which the community works together in a mutually supportive way and utilizes indigenous technologies. In some instances the purpose of a project seems less for the benefit of the villagers than for a foreign project worker who wishes to win approval and forgiveness from villagers unsullied by the evils of modernity. Much of the driving force behind such projects comes from a gut reaction against the alienation and consumerism of modern society rather than any cold political analysis. The notion of the Noble Savage has a long history and that of Utopia even longer. These ideas have led to rebellious but relatively wealthy individuals establishing model communities in remote places and populating them with 'innocents'. Many aid workers appear to be using the villages of the Third World as laboratories for their own Utopian social experiments.

Those who do not share the relevant faith condemn the imposition of these religious and political agendas as imperialistic. But to some within the faith the cause is righteous and any means to the desired end are acceptable. The believers see themselves as failing in their duty if they do not use the power at their disposal to spread the Good News, whether of God, Marx, Mammon or Guia.

At certain historical periods the fundamentalists have been in the majority. In Victorian England, the righteousness of the missionary's calling was largely unquestioned. For much of the twentieth century, the Marxists have been equally sure that they are right. In recent years, for many of us, such firmness of faith has itself become the object of suspicion. A tentative belief in a more or less vague cause is countered by doubts as to the acceptability of the means. To what extent is it legitimate for a donor to use his financial might to override a community's desire for concrete blocks for the sake of 'appropriate' local materials? Lately, a new faith has emerged which has come to have the potency of Christianity, Islam, Capitalism and Marxism before it. The Green imperative offers a new certainty; a cause which the believers, like their predecessors, *know* is correct.

In the past, powerful international institutions like the World Bank have been labelled, derogatorily, as imperialist. Today, the environmental lobby is having some success in persuading such institutions to use their muscle to impose Green policies on governments. Such methods are imperialist. It seems that imperialism, the old enemy, was never really the problem. The issue is about determining the justice of the cause. Self-determination is good but the survival of the planet is better. In village-scale projects this newly emerging certainty is leading to a similar confidence in the imposition of criteria. Earlier we had known that eating people was wrong, we now also know that slash and burn agriculture, excessive use of agricultural chemicals and high-energy building materials are wrong.

The slogan 'bottom-up development' may be a useful and attractive fund-raising message, but the aid intervener should not be taken in by his own rhetoric. In practice, the villagers are almost never taken to be the sole arbiters of what is good nor, in theory, should they be. For the intervener, blind acquiescence to the villagers' demands would mean relinquishing his or her moral obligation to determine what it is he or she believes to be good and to fight for it. I am not suggesting that aid institutions should revert to insensitive, top-down, centrally-controlled projects which are imposed on an unwilling population. Rather, that the intervener should cease any covert imposition of values behind the jargon of empowerment and openly declare his beliefs and argue for them. The hazy principle that 'the people' are always right is not only irrational, it is also patronizing. The intervener can best show his respect for the villagers by treating them as consenting responsible adults with whom one can argue, disagree and negotiate.

To arrive at a policy an intervener has a range of tools at his disposal: in-depth discussion with the villagers, anthropological studies, experience of other projects and consideration of global criteria. At some point a decision has to be made. The decision is never final – just a best bet – but it is none the less a decision. If the intervener's conclusions are significantly different from those of the villager, the intervener has three possible courses of action: to accept defeat, use power to impose a solution or to use persuasion. All three strategies have their place. The imposition of a solution is clearly a last resort and in many instances is not possible; a health programme cannot force people to wash their hands. If the intended beneficiary, the villager, is free to say no and the intervener considers the issue to be important then the only strategy left is persuasion.

PERSUASION

Systematic persuasion, whether in the form of friendly conversation or slick video technology, is synonymous with advertising and marketing. The outrage invoked by these two words was touched on earlier. To many, marketing suggests the generation of desires for unnecessary products which solve spurious problems in order to make profits for parasitic financiers; or it suggests the manipulation of people's weaknesses to make them act in ways other than they would wish. Clearly, marketing techniques are often used to these ends, but not exclusively so. There are plenty of examples of government-sponsored advertising in the interests of the public good rather than private profit. The techniques used to promote safer driving, discourage smoking and help people avoid AIDS are all culled from commercial advertising. Within reason, it is not the techniques which are good or bad so much as the message. While the objective of commercial advertising is to maximize profits for shareholders, the objective of development education is to maximize the overall benefits to the villagers. This objective places a burden of honesty on the intervener not only to emphasize the attractions of an idea but also to spell out its limitations.

Paulo Freire described his influential work with shanty town dwellers in Brazil as 'consciousness raising' in order to help people understand their situation (Freire 1972: 75–118). The task he faced was not unlike that the designers of an advertising campaign to stop smoking confront. He was trying to make people aware of links between superficially unconnected issues, in his case the wealth of the capitalist owners and the poverty and oppression of the people he was working with. The moral basis for the concept of consciousness raising has, however, been questioned:

> A crucial assumption of the concept is that lower-class people do not understand their own situation, that they are in need of enlightenment on the matter, and that this service can be provided by selected higher class individuals.
>
> (Berger 1974: 148)

The same assumption underlies all aid. However honest and sincere the intent, the purpose of development education is to change people's attitudes and knowledge in ways which we consider to be for their good. Whether the activity is called public service advertising, propaganda, technical aid, brain-washing, development education, consciousness raising or life-style marketing, the activity is essentially the same and it is inescapably paternalistic – we believe that we are enlightened and that we have truths to pass on to them, the unenlightened. If one accepts the justification for technical aid given in the introduction of this book then one has to accept the practice of the art and craft of persuasion.

There are many available techniques for persuasion, from posters and videos through to participatory workshops and the use of drama. The intention here is not to conduct a review of these techniques but simply to appeal for clarity in defining the objective of any act of persuasion. What is the Big Idea for the intervener and what is the Big Idea for the villager? Frequently development programmes are trying to persuade villagers to adopt solutions to problems they are unaware that they have. Before one can sell a solution it is often necessary to sell a problem. That a person cannot diagnose a disease does not mean they are not suffering from it. Ignorance of the consequences of dirty water does not make it any less of a threat. This is why it is not enough simply to listen to people's problems. A person who can clearly articulate his or her problem is three quarters of the way to solving it – which was precisely Freire's point. The argument which Freire used for liberation also applies to diarrhoea. The difficult part is framing the problem in the right way. If one is aware of the link between defecating near water sources and diarrhoea then the task of identifying the technological solution is trivial.

Our problem statements tend to carry within them pointers towards the kind of solution we expect and effectively to exclude others. Often, the problem statements of intervener and villager point in contrary directions. The three examples in Table 3 illustrate that the same objective problem may be formulated in two different ways which lead towards different kinds of largely pre-determined solutions.

The prime purpose of persuasion is not necessarily to bring about a congruence in problem statements between intervener and villager so much as problem statements which can be resolved by means of the same solution. If the problem statement of the villager is specifically 'I do not have a water closet' then a composting latrine – however nicely presented – will not satisfy that clearly specified need. The problem stated as 'I do not have sanitation' could be met by a latrine but, as was suggested in Chapter 6, such a statement is

Table 3 Hypothetical intervener and villager problem statements

	Intervener		Villager	
solution	*← problem statement*	*problem statement →*		*solution*
Hygiene and sanitation	The have parasite infestation due to lack of hygienic sanitation	My children are dying because I cannot afford the medicines		Money and medicines
Water catchments and ground cover	Their crops are dying because the rain water runs off and evaporates too quickly	My crops are dying because they are receiving too little water		Irrigation
Pivot door	Their doors are falling out because the hinges are placing too much stress on the walls	The door of my house is falling out because the earth walls are not strong enough		Concrete block walls

generally unlikely. For many, the latrine has no place either in the traditional, understood, life-style or in the aspired-for modern one.

The problem is not exclusively one of promoting the objective attributes of a given product but of demonstrating how the adoption of the product can help to meet higher-level targets on a broader agenda aimed at attaining a certain life-style. If there is a higher-level objective for a modern bathroom then the possibility exists for presenting an option which is recognizably a modern bathroom but which contains the novel feature of a composting latrine, particularly if the money saved can contribute to another project on the broader agenda. In Chapter 1 it was suggested that knowledge advances in discrete, testable steps. The same, I suggest, is true for aspirations. The commercial advertiser will show respectable, recognizable types of people in equally respectable settings with one novel item, the new product. The target audience will be invited to evaluate the product in terms of its proposed setting. If the dreamt-for solution is a water closet in an urban bathroom then the first demonstration composting latrine, with its thatch roof and garden, offered a picture in which virtually every element conflicted with the dream. The second latrine provided an already aspired-for context in which a single novel idea could be presented.

THE PRINCIPLE OF MULTIPLE AGENDAS

Accept that the intended beneficiary and the intervener may regard different aspects of a technology as being advantageous or unacceptable. Recognize that, on occasions, it will be necessary to try and persuade a villager to see his or her problems in a different way.

The Principle of Multiple Agendas suggests that the intervener should ask: 'what am I trying to achieve and what is the villager and everyone else involved trying to achieve?' The intervener should be more relaxed on the subject of Appropriate Technology than many aid institutions currently are. The object should not to be to achieve unanimity regarding some perfect, ideologically sound solution but rather a technology which everyone involved can live with. It advocates being honest about a technology's weaknesses while stressing those aspects which the villagers are most interested in. The intervener will need to:

• Clarify what his or her general and specific agendas really are.

The social scientist can help to:

- Determine, as far as is possible, what the relevant parts of the villager's agenda are.
- Determine who else may be involved in a particular situation and how their agenda may differ.

The challenge for the technology designer is:

- Where possible, to identify a technology which satisfies the agendas of all those involved.
- Where this is not possible, to identify how the agenda of the villager or anybody else may have to alter for a given solution to become acceptable.

CONCLUSIONS

> In all technical change, even when it seems to be concerned
> with tools, machines and other impersonal objects, the indivi-
> dual person is both the recipient of change and the mediator
> or agent of change.
>
> (Margaret Mead 1955: 288)

Ten years before writing this book, community participation was
already being described as 'the single most-written-about issue in
the field of rural development' (Chaufan 1983: 9). Advocacy of
participatory techniques can be found in the writings of progressive
colonial administrators in the 1920s and 1930s (Hardiman 1986: 54).
For a generation, participation has been a mainstay of academic
writing and teaching about development. By the 1970s the policy
statements of the major international donors and implementing
agencies all emphasized the importance of participation.

It is time to stop simply reiterating the cry for community
participation. That was yesterday's battle. Certainly, despite all the
rhetoric, participation often does not happen. But this failure will
not be rectified by yet more books and speeches calling for
participation. The challenge is now to get beyond the general
principle and determine the practicalities of how participation fits
into a larger picture of effective aid for just and sustainable
development.

A ROLE FOR PARTICIPATION

In the introduction of this book it was suggested that participation
can function either as a tool or as a goal. Participation as a goal in
itself needs to be approached with care and intellectual rigour. One

author observes that 'it is a myth to assume that everybody wants to be actively involved in decision-making' (Hardiman 1986: 65; see also Midgley 1986: 36). Another concludes that the ideal of 'full participation' in which everybody participates in every decision would 'constitute a nightmare comparable to unending sleeplessness' (Berger 1977: xvii). In many circumstances, the very ideas of community participation and democracy can be externally imposed concepts based on western ideology rather than local practice (Midgley 1986: 152; see also Chambers 1983: 144). Which is more ideologically unsound: the tolerance of traditional modes of repression or the attempt to impose a democratic system?

The romantic view of participation focuses on the personal fulfilment which it can bring. But true participation is about power, and the exercise of power is politics. This kind of participation inevitably becomes simply a manifestation of a broader political process. If community participation is to be encouraged as an end in itself, it cannot be divorced from its political context and consequences. When power is challenged it generally provokes retaliation. Any aid programme intending to use the lives of others as pawns in a political struggle must be certain of its facts and morals and must be prepared to see the struggle through. These observations do not suggest that community participation as an objective of aid should necessarily be discouraged but rather that it should be recognized for what it is – an externally motivated political act (Berger 1974: 148). It might be more useful to call such notions of community participation 'democracy', and debate and fight for its merits in the appropriate political context; local, national or global.

Whatever the political motivation for a project, the day-to-day practice of aid demands tangible products. Each step needs a specific goal: increased literacy, healthier children, employment opportunities for women or a new water supply. In order to achieve these local goals, irrespective of any broader objective, the function of community participation is that of a tool to achieve a given task.

Some projects require community participation to carry out such tasks as digging a ditch or building a bridge. But this participation is purely one of responding to outside demands – it is not so much participation as passive collaboration with *our* intervention. It would be best if such collaboration was never referred to as community participation, but rather as something quite distinct, such as the community's contribution. Other projects involve community committees to determine which families should benefit or which individuals should be employed in community enterprises. Although this is participation in a more real sense, it is still participation within the limits determined by the project. It is community management rather than self-determination.

Community contribution and community management certainly have a place in the lexicon of aid, but the real interest lies in finding mechanisms through which the community can influence the nature of the aid which is being offered to them. But should not 'influence' be replaced by 'control'? My contention is that talk of the community control of aid is, at best, misleading. Aid is predicated on an 'us' and a 'them' where 'we' have resources which 'they' want. Mechanisms can certainly be introduced in which community representatives control the detailed disbursement of funds or whatever other resources are on offer. But the real control of whether or not a given community will be aided inevitably lies with those who hold the purse strings. Those strings can be handed over to 'the community', but then someone has to decide what constitutes a community and to which community the strings can be handed. True community control will only exist when aid ceases to exist, because the disparities of wealth which give rise to aid will have vanished. We are a long way from that point.

Thus in my view, the most honest and fruitful focus for community participation in the aid process is in influencing the nature of the aid offered in order that the aid can be made more acceptable and relevant to the intended beneficiaries.

ACTION AND REFLECTION

While the words 'participation', 'empowerment' and 'sustainability' are currently in fashion, the phrase 'transfer of technology' is now discredited in academic circles. Transfer of technology is regarded as an imperialist notion implying the imposition of external ideas and values and an undervaluing of indigenous knowledge. Yet it is the contention of this book that we should learn to transfer technology more successfully since:

- Aid means the transfer of material or intellectual resources from rich to poor.
- The arguments which have been presented in this book concerning the transfer of physical technologies also apply to the transfer

161

of less tangible goods such as improved teaching, promotion or organizational skills. The principles are simply easier to grasp when applied to physical things.

Freire may have been interested in the liberation of the oppressed but he recognized that his agenda could best be met by offering a tangible skill. Literacy is not only a powerful tool through which one can learn to understand one's situation, it also has a name, it makes sense in the villagers' own terms as a means to get services and employment, and it is a universally respected skill. If Freire, like many before him, had simply tried to lecture to the people about liberation he probably would not have had much success. Literacy was a technique which opened the door to a process of community-based development.

It is, in my opinion, mistaken to suggest that there are two opposing models: community-based participatory development versus the transfer of technology model. The two are not rival processes but rather they correspond to two necessary and complementary stages of a single cyclical aid process:

- **Action** The action, or implementation, stage involves the transfer of some resource: materials, money, knowledge or values. It is characterized by the intervener pursuing a course of action with single-minded determination because he or she believes it to be right. It is about 'us' transferring something to 'them'.
- **Reflection** The process of reflection can be broken up into the stages of evaluation, problem definition and the trial of proto-types. At these moments in the project cycle, the concern is how to assess and improve the quality and content of the action. The reflective process is greatly enhanced by learning from the intended beneficiaries – 'they' transfer knowledge to 'us'.

The immediate object of aid is the action component. That is the point – the only point – at which the intended beneficiaries stand to gain. Action does not automatically mean the introduction of ingenious technical gadgets. It may mean a programme of meetings to raise consciousness with a view to precipitating the revolution; whether political, agricultural or sexual. The common factor linking together different types of action is that there is an intervener who is trying to get a group of people to change in a given way through the introduction of ideas or things. In a development programme each action creates the context for subsequent reflection and reflection suggests further action.

Before action is taken the intervener has chosen a line to pursue. But how is that initial decision made? Where the problem definition and the development of the response are worked out in offices remote from the intended beneficiaries the assessment of need is likely to be poor and the consequent action misdirected and ineffectual or even damaging. In such cases the fault lies not in the idea of transferring resources but rather in the ill-informed judgement of the project's designers.

To improve the quality of the action it is essential to listen to the intended beneficiaries during problem definition, trials of prototypes and evaluation. But, however important it is, discussion with the community is only one of several techniques which can be brought to bear on these tasks. The evidence of passive observation, previous experience and scientific knowledge exists and is too valuable to waste. And, as discussed in Chapter 3, listening to the people is not always a straightforward exercise.

The focus of much contemporary writing on development has been on participatory research (see for example McCracken *et al.* 1988; Chambers *et al.* 1989). This is a highly important field which needs to be nurtured and absorbed into the routine practice of aid. However, while it is a necessary part of an aid project it is not sufficient in itself. There is a danger in creating the illusion that participatory research is all there is to ideologically sound aid. If the process does not get beyond the stage of problem definition then it is not a process of aid but rather of academic anthropology or development tourism.

By its nature, participatory research can, on occasion, result in direct benefits for the community. But in these instances it is not so much participatory research as the activity of 'consciousness-raising' in Paulo Freire's sense – of helping people to understand their circumstances. Participatory research workshops can only function with small groups of people. The problems are vast. If aid is intended to benefit large numbers of people the participants in research workshops and the like can only ever be small samples of the intended beneficiaries. Participatory research only makes sense as a data gathering component in a larger aid process which includes wide-scale action.

By placing a box around participatory research as a discipline for development academics, unconnected with the dirty work of transferring resources, there is a real danger of reinforcing the divisions which so often separate reflection from action. In a good project, the action/reflection cycle will be re-enacted many times, not just once. To achieve this happy state the research activities must be intimately linked to the activity of the transfer of resources. The method and content of such a transfer should emerge from the community-based research, but equally, to be useful, the research should be informed by an understanding of the dynamics of the transfer of knowledge and resources.

Though the object is to change *their* process – the development process – when we talk about the aid process we are talking about *our* process. Participatory research fulfils the same function as the market research exercises carried out by commercial companies. As with development research, market research can consist of crude questionnaires or sophisticated 'focus groups' which allow for a more profound examination of people's attitudes and aspirations. But whatever the source and validity of the research its only role is to produce data and recommendations for the policy makers. After twenty years at the top of the rhetoric mountain, community participation needs to come of age and become a focussed tool for executing the specific task of research to inform action.

The task of the policy maker is to weigh up the evidence, including that from participatory research, and decide on a course of action as a calculated risk. As stressed in Chapter 8, on occasions the action chosen will be at odds with the expressed priorities of the community, and any course of action can never be more than the best bet of the moment. But once an action has been chosen the emphasis shifts from the community's participation in our process to the intervener's attempt at participation in the community's process.

Community participation in our decision-making process is a false horizon – it is a necessary step but it is not a goal in itself – we must advance our thinking beyond it. Community participation in our process is only important in so far as it can help us to improve the quality of our intervention in their process. It is their process which matters.

THE THREE R'S

Any programme of technical aid is only as good as its field workers. Field workers are often under-rewarded and their knowledge undervalued by their employers. Partly this is a consequence of aid institutions being dominated by an academic urban elite which has never had to deal on a day-to-day basis with problems in a village, a shanty town or on a factory floor. But it is also due to the fact that the parties lack a common vocabulary for discussing their common problems. This book has attempted to establish a few grammatical principles for such a vocabulary.

The principles suggested here may seem esoteric and alien to the realities of development aid and the budget lines of project proposals. But much development aid fails precisely because, despite all the talk about the community, it does not fully consider what development means for the individual. Development is not just about increased wealth. It means change; changes in behaviour, aspirations, and in the way in which one understands the world around one. Without an understanding of how and why people change, knowledge of the mechanics and high finance of technical aid will be of little use. Only by understanding the fine grain of change in the village and the shanty town can large-scale policies benefit individual's lives.

In this book it is suggested that three basic questions summarize the process by which a person assesses a new idea:

- **Does it make sense?** Is the idea *reasonable* in terms of the intended beneficiary's own rationale?
- **What is it?** Can the idea even be *recognized* – does it have a name and are its limits clearly defined?
- **Is it worthy of me?** Is the idea *respectable* – Is it something which 'people like us' do?

From a consideration of these questions emerge eight principles for the design of interventions. But observance of the principles is no substitute for either a good idea or a committed, intelligent and sensitive field worker. Principles cannot generate good ideas, they can only help in the processes of evaluating and choosing between options. While a good worker or policy maker already has an intuitive feel for such principles without any need to articulate them, a formal list can still assist in discussing proposals and helping others to draw out the accumulated experience of the good field worker.

The criteria of change suggested here – *reasonableness, recognizability* and *respectability* – apply just as much to university-trained engineers or doctors as they do to village farmers or street vendors.

Many intelligent, well-intentioned people involved in aid are ignoring key problems simply because they have not recognized them as problems. For them, as for the farmer, the criteria of respectability is vital. Development workers need to feel that they are tackling respectable problems that can be subjected to professional analysis and evaluation. In the same way that a villager needs to feel that 'people like us' would use this technology, so too does the development worker, at whatever level, need to believe that this development activity is what 'people like us' do.

Aid is about stimulating change. To be more likely to succeed aid must be change-like – it must share characteristics with the massive and powerful processes of change which are going on all the time. In summary the three basic principles of change-like aid are:

Reasonable

For a new idea to be adopted it must make sense in terms of the intended user's own rationale. It must be clear what aspect of the idea is new within the context of existing knowledge and it must fit into the understood social fabric of responsibilities and skills. In order to understand what other people will consider reasonable it is necessary to find ways of learning about the criteria, knowledge and priorities of others.

However painfully obvious the principle of reasonableness may seem, it is clear that many projects being written and implemented today fail to ask the question, 'will this make sense to the intended beneficiaries?' Notions of what is reasonable are highly subjective and dependent on what is already believed. What is self-evidently reasonable to someone from a western industrialized country or a Third World capital city may seem either mysterious or mistaken to someone from a different culture.

Many technical aid projects fail because one person is trying to impose his idea of reasonableness onto other people: the frustrated bull-like engineer who, however much he shouts, cannot get his counterparts to install the machinery *his* way, unaware that they are only too aware of future problems relating to spare parts and fuel; the agricultural expert cajoling people into borrowing money to buy 'improved' high-yield pigs when they have nothing to feed them with; the committed young doctor advocating birth control to people who in old age will depend on children to support them.

166

Recognizable

Before anyone can evaluate and adopt an idea he or she has to understand what it is. Often an intervener and the intended beneficiary are talking at cross-purposes with the intervener moving on to a consideration of details before the basic idea has been grasped. A new idea must fit into a person's existing structure of knowledge and have a name. Educational materials which are used to try and explain the idea must be comprehensible to the intended beneficiary.

Paradoxically, the principle of recognizability is often the hardest to get across. Anyone who has tried explaining a new idea to a child knows the frustration on both sides when the child cannot comprehend an idea which to the adult seems self-evident. The real problem is that the adult does not know what it is that he himself knows. When a villager tries to explain something to an outsider his frustration can be similar and vice versa.

The training of development workers – from the local field worker to the university-trained specialist – seems almost designed to obscure communication. Often, local health workers are trained to use medical jargon in preference to the day-to-day terminology of the people they are meant to serve. Only by understanding the way in which the intended beneficiaries think about and describe the world around them can one design an intervention with a recognized purpose and place.

Respectable

People want self-respect. Whatever their economic circumstances they still want to be respected by their peers. People only adopt ideas which they consider proper to people in their circumstances; suitable for 'people like us'. Increasingly, in a rapidly changing world this means being modern. An intervener must understand who or what it is that defines respectability. Often the criteria of interveners and the target population are at odds. The intervener must search for solutions which are acceptable to all concerned and, if need be, begin by example and persuasion to help redefine what is considered respectable.

The principle of respectability is the most frequently overlooked. Despite the wide range of cultures, we all have certain characteristics

167

Figure 18 An abandoned dream house
The sight of abandoned, half-completed concrete houses is becoming
increasingly common as people's aspirations and expectations get further
out of step with their resources. People need more resources and more
achievable dreams.

in common and one is the need for self-respect. Although the
elements which define respectability vary enormously, the underlying principle is constant. Aid programmes which focus on 'objective' or 'basic' needs and ignore the social acceptability of their
offers are likely to fail.

168

THE GLOBAL VILLAGE

Even though 'we', the interveners with surplus resources to donate, are inevitably distinct from 'them', the intended beneficiaries, we are linked in ways other than the contracts of aid. We occupy the same planet, which with telecommunication and travel is shrinking fast. In the global village the intervening classes are becoming role-models of modernity which, as has been suggested, is increasingly synonymous with respectability.

In recent years 'sustainability' has come to occupy a place in the vocabulary of aid on a par with 'community participation'. If the courses of action proposed by aid must be sustainable and if the suggested options must seem modern to be acceptable then it follows that modernity itself must take a form which is sustainable. Clearly, the way of life led by many in the intervening classes – you and I – is not sustainable; particularly if extended to everyone on the planet.

However sophisticated and well formulated future aid interventions become they will all count for nothing if the model to which we all aspire is not sustainable. The greatest contribution which we, the intervening classes of both the advanced industrial countries and the cities of the Third World, can make is to minimize hypocrisy and, as far as is possible, live and be seen to live the way of life which we recommend to our poorer neighbours in the global village.

REFERENCES

We approach everything in the light of a preconceived theory.
So also a book. As a consequence one is liable to pick out
those things which one either likes or dislikes or which one
wants for other reasons to find in a book.

(Karl Popper 1970: 51)

Alexander, C. (1964) *Notes on the Synthesis of Form*, Cambridge, Mass.:
Harvard University Press, (references for 9th edition 1977)

Aysan, Y. and Davis, I. (Eds) (1992) *Disasters and the Small Dwelling*,
London: James and James

Barley, N. (1983) *The Innocent Anthropologist*, London: British Museum
Publications (page references refer to the 1986 Penguin edition)

Berger, P. L. (1974) *Pyramids of Sacrifice: Political Ethics and Social
Change*, New York: Basic Books

Berger, P. L. (1977) *Facing Up To Modernity*, New York: Basic Books

Berger, P. L., Berger, B. and Kellner, H. (1974) *The Homeless Mind*,
New York: Basic Books

Berlin, B. and Kay, P. (1969) *Basic Color Terms*, Berkeley: University of
California Press

Birkinshaw, M. (1989) *Social Marketing for Health*, Geneva: WHO

Brandberg, B. (1985) *The Latrine Project, Mozambique*, International
Development Research Centre manuscript report No.58 (revision E):
Ottawa, Canada: IDRC

Brokensha, D. W., Warren, D. M. and Oswald, W. (Eds) (1980) *Indigenous
Systems of Knowledge and Development*, Washington, DC: University
Press of America

Burgess, R. (1982) 'Self-help housing advocacy: a curious form of
radicalism, a critique of the work of John F. C. Turner', in P. M. Ward
(1982) *Self-Help Housing: A Critique*, London: Mansell, 56–98

Carr, M. (ed.) (1985) *The AT Reader*, London: I.T. Publications

Chambers, R. (1983) *Rural Development: Putting the Last First*, London:
Longman

Chambers, R., Pacey, A. and Thrupp, L. A. (1989) *Farmer First: Farmer
Innovation and Agricultural Research*, London: Intermediate Technology
Publications

Chaufan, S. K. (1983) *Who Puts the Water in the Taps?*, London: Earthscan

Cobbett, W. (1830) *Rural Rides*, republished 1967, London: Penguin

Conroy, C. and Litvinoff, M. (1988) *The Greening of Aid*, London: Earthscan

Conway, G. R. (1989) 'Diagrams for farmers', in Chambers, Pacey and Thrupp (1989) 77–86

Conway, G. R., Husain, T., Alam, Z. and Mian, M. A. (1988) 'Rapid rural appraisal for sustainable development, the northern areas of Pakistan', in Conroy and Litvinoff (1988) 178–85

Dudley, E. (1992) 'Disaster aid: equity first', in Aysan and Davis (1992)

Dudley, E. and Haaland, A. (1993) *Communicating Technical Ideas*, London: Intermediate Technology Publications

Freire, P. (1972) *Pedagogy of the Oppressed*, London: Sheed and Ward

Fuglesang, A. (1982) *About Understanding – Ideas and Observations on Cross-Cultural Communication*, Uppsala, Sweden: Dag Hammerskjold Foundation

Gupta, A.K. (1989) 'Scientists' views of farmers' practices in India: barriers to effective interaction', in Chambers, Pacey and Thrupp 1989: 24–31

Haaland, A. (1980) *Communication, Social Change and Attitudes*, unpublished paper prepared for the staff of the Bhaktapur Development Project, Nepal

Haaland, A. (1984) *Pre-testing Communication Materials* – A Manual for Trainers and Supervisors, Burma: UNICEF

Haaland, A. and Fussell, D. (1976) *Communicating with Pictures in Nepal*, Kathmandu: UNICEF

Hall, A. (1986) 'Education, schooling and participation', in Midgley (1986) 70–86

Hardiman, M. (1986) 'People's involvement in health and medical care', in Midgley (1986) 45–69

Harrison, P. (1987) *The Greening of Africa*, London: Paladin

Hayter, T. (1981) *The Creation of World Poverty: An Alternative View to the Brandt Report*, London: Pluto Press

Hayter, T. (1987) 'Poverty amid plenty: the need for more democracy', pp. 99–104, in Building and Social Housing Foundation (BSHF) (1987) *Home Above All*, Coalville: BSHF

Holt, J. (1983) *How Children Learn, Revised and Expanded Edition*, Harmondsworth: Penguin

Honadle, G. (1979) quoted in McCraken *et al.* (1988) 21

Howes, M. (1979) 'The use of indigenous technical knowledge in development', *IDS Bulletin*, 10, 2, pp. 5–11

Ignatieff, M. (1984) *The Needs of Strangers*, London: Chatto & Windus, London: The Hogarth Press

ITDG (Intermediate Technology Development Group) (1967) *Tools for Progress*, London: ITDG

Jéquier, N. (Ed.) (1976) *Appropriate Technology: Problems and Promises*, Paris: Development Centre of the Organisation for Economic Co-operation and Development

Lamug, C. B. (1989) 'Community appraisal among upland farmers', in Chambers, Pacey, and Thrupp (1989) 73–7

Lerner, D. (1958) *The Passing of Traditional Society: Modernizing the Middle East*, New York: Free Press

Leslie, J. (1987) 'Think before you build, experiences after the Yemen earthquake', *Open House International*, 12, 3

Levi-Strauss, C. (1961) *World on the Wane*, London: J. Russell (originally *Tristes Tropiques*, Paris 1955)

Linear, M. (1985) *Zapping the Third World: The Disaster of Development Aid*, London: Pluto Press

McBean (1989) *Re-Thinking Visual Literacy, Helping Pre-Literates Learn*, Kathmandu: UNICEF

McCracken, J. A., Pretty, J. N. and Conway, G. R. (1988) *An Introduction to Rapid Rural Appraisal for Agricultural Development*, London: International Institute for Environment and Development

Mathema, S. B. and Galt, D. L. (1989) 'Appraisal by group trek', in Chambers, Pacey and Thrupp (1989) 68–73

Mead, M. (Ed.) (1955) *Mental Health Implications of Technical Change*, New York: UNESCO

Midgley, J. (Ed.) (1986) *Community Participation, Social Development and The State*, London: Methuen

Murdoch, P. (1981) 'Vernacular house form in Ladakh', in Toffin (1981) 261–78

Oxfam (1987) *Water Harvesting in the Yatenga*, Information sheet Burkina Faso 93, Oxford: Oxfam

Peirce, C. S. (1966) *Selected Writings*, Ed. P. P. Weiner, New York: Dover

Popper, K. (1970) 'Normal science and its dangers', in Lakatos, I. and Musgrave, A. (Eds), *Criticism and the Growth of Knowledge*, Cambridge: Cambridge University Press

Rapoport, A. (1969) *House, Form, and Culture*, Englewood Cliffs: Prentice-Hall

Reij, C. (1988) 'Soil and water conservation in Yatenga, Burkina Faso', in Conroy and Litvinoff (1988) 74–7

Richards, P. (1980) 'Community environmental knowledge in African rural development', in Brokensha *et al.* (1980) 183–203

Richardson, C. (1893) quoted in *Essays on Rural Hygiene* by Poore, G. V. (1893) Longmans, Green and Co: London

Rondinelli, D. A. (1983) *Development Projects as Policy Experiments*, London: Routledge

Rybczynski, W. (1980) *Paper Heroes*, Dorchester: Prism Press

Schumacher, E. F. (1973) *Small is Beautiful*, London: Blond & Briggs (references for the Abacus/Sphere Book 9th edition 1978)

Schutz, A. (1964) *Collected Papers, volume I*, The Hague: Martinus Nijhoff

Soedjarwo, A. (1983) 'Owner-built birth control', in Carr (1985) 232–3

Spence, R. J. S. (1987) 'Soil-cement blocks: selection of soils and cement content', *Building Technical File*, 19

Thomas, J. W. (1975) 'Irrigation tube-wells in Bangladesh', in Carr (1985) 125–30

Tisa, B. (1984) 'A village postman's story', *Appropriate Technology for Health*, newsletter 14–15, p. 25, Geneva: WHO

REFERENCES

Toffin, G. (Ed) (1981) *L'Homme et la Maison en Himalaya*, Paris: Editions de Centre National de la Recherche Scientifique (CNRS)

Turnbull, C. (1972) *Mountain People*, London: Jonathan Cape, (references for the Triad/Paladin edition, 1984)

Turner, J. F. C. (1976) *Housing by People*, London: Marion Boyars

Turner, J. F. C. and Fichter, R. (Ed.) (1972) *Freedom to Build*, New York: Macmillan

Werner, D. (1977) *Where There is No Doctor*, London: Hesperian Foundation/MacMillans

Willoughby, K. W. (1990) *Technology Choice: A Critique of the Appropriate Technology Movement*, London: Intermediate Technology Publications